QRP Basics

Third edition

by Rev George Dobbs, G3RJV
Edited by Professor Roger Green, M0RJG

Radio Society of Great Britain

Published by the Radio Society of Great Britain, 3 Abbey Court, Fraser Road, Priory Business Park, Bedford MK44 3WH. Tel: 01234 832700.
Web: www.rsgb.org

First published 2003
Second edition 2012
Third edition 2021
Reprinted Digitally 2023 onwards

ISBN: 9781 9139 9509 6

Sub-editing: Ed Durrant, DD5LP

Cover design: Kevin Williams, M6CYB

Production: Mark Allgar, M1MPA

Typography and design: Mark Pressland

Printed in Great Britain by 4Edge Ltd. of Hockley, Essex

Any amendments or updates to this book can be found at: www.rsgb.org/booksextra

Contents

Chapter 1

QRP: A Personal Introduction

George Dobbs, G3RJV

George Dobbs continues to be an inspiration for all those in the amateur radio community who have any interest in QRP, that is, low power radio communications. In his early days, his own personal inspiration was Doug DeMaw, W1CER, as he described below, but George followed on from there and spread the gospel – in more ways than one! As well as being a substantial aficionado in amateur radio, in his youth he also felt a strong religious calling and became a priest. He will tell more about his history in his own words, but, suffice to say, I feel very honoured to have been asked by the RSGB to review and edit his book, and to provide further material in keeping with George's interests. This has been a challenge, as George will always be regarded as the expert, and I feel that one must tread very warily when making changes or additions to his work. So, what follows is largely George's work from previous editions of this book, revised only where approriate.

Prof. Roger J. Green, M0RJG

Amateur radio has a fine reputation for being a life-long hobby. I have been a radio amateur for 40 years, but there are times, even in such a compulsive hobby, when the interest wanes. One such time for me was at the beginning of the 1970s. I had just moved home for the third time in a few years and a whole lot of other interesting things were happening in my life. So, when I arrived at the new house I did not even bother to unpack the boxes that contained my amateur radio equipment. Neither did I bother with the usual radio amateur tendencies when moving in, of identifying a suitable room for a radio shack and inspecting the garden space for suitable antenna erection. There were other things to do.

After I had been in the new location for a couple of weeks, a member of the local amateur radio club called to see me with an invitation to a meeting. I think he sensed my lack of interest because he returned a few days later looking for bargains. We opened the boxes and pondered over the array of equipment and by the end of the evening he went away with everything I had . . . as I recall it, at a bargain price.

For a few months I had no amateur radio equipment at all. But gradually the old longing came back and I began to look around at ideas for building an amateur radio station again. Most of what I had sold was quite old, most of it was homemade and all of it was valve technology. I thought it about time to launch in to solid state technology and to find out about, and use, transistors.

It was about this time that Heathkit produced a simple solid-state transceiver called the HW-7. This was a direct conversion transceiver for the 40, 20 and 15 metre bands (7MHz, 14MHz and 21MHz) that had a transmitter output power of some 2W (on a good day). But it was simple, it was a self-contained amateur radio station and the kit was relatively cheap. At least, cheap enough

The author, George Dobbs (SK), G3RJV, in the operating corner of his workbench

for me to buy. The Heath Company of Benton, Michigan had a fine tradition of producing kits with quality components and very comprehensive construction instructions.

True to the Heath tradition, the kit went together easily but I had real doubts about its ability. There seemed to be too few parts in the circuit to make up a complete transceiver. I had not used direct con- version as a technique before and it seemed too simple to be viable on the amateur bands. So, when the HW-7 was completed, I was anxious to see if it would serve me as a station.

The Heathkit HW-7 transceiver

Almost all of my former station was gone but I managed to find an old car battery to provide the 12 volts required to power the transceiver but I did not have any antennas. I decided to make a 20 metre (14MHz) dipole. After cutting the two elements of the dipole, I realised that I did not have any co-axial cable to use as a feeder line, so I settled for a directly fed dipole. The two sections of wire were drawing-pinned (thumb-tacked) along a picture rail and met at a table in a bay window of a ground floor room. One half was connected to the antenna socket and the other was attached to the case of the transceiver. I had a dipole antenna for 20 metres, albeit very low off the ground.

Switching on the HW-7 and tuning it across the 20-metre band was very gratifying. The simple circuit was picking up a lot of stations. But would it transmit? Yet another problem – I had sold all my Morse keys. The answer was to attach a clip lead to the 'key' socket and tap the other end of the lead on the back of the case. Each time the lead touched the back of the case, the needle on the front panel shot across the scale. It did appear to be transmitting.

Tuning across the band, I heard a Swedish station calling "CQ". On impulse, I decided to try calling him by tapping the clip lead on the case of the transceiver. To my amazement, he replied. The amazement turned to shock when he gave me an RST report of 569. There I was with an output of about 2W feeding two pieces of wire in a downstairs room and this man in Sweden was claiming that I was signal strength 6! I would not have been unhappy about that report for my old transmitter, which gave about 40W output from a valve. Up to that time, I had

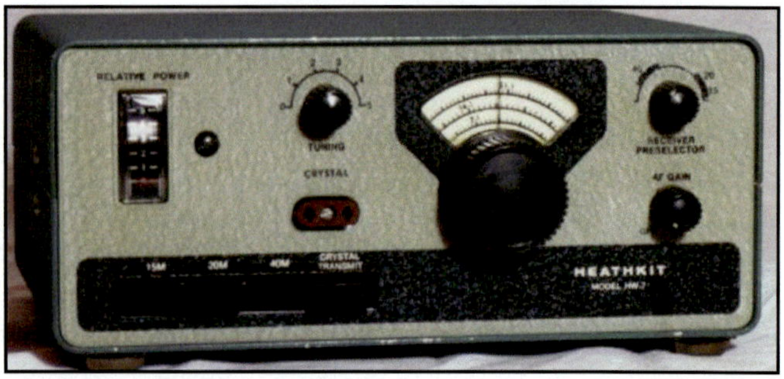

Heathkit HW-7

never really thought about the implications of power level from a transmitter. I will say more about that later in this book. My success spurred me on to install better outside antennas and I began to gather a respectable countries-worked score with my little transceiver. It did not cover the 80 metre (3.5MHz) band, which had been a favourite of mine, and I was encouraged by the simplicity of the technology to build my own transceiver for that band. Copying most of the Heath circuitry, I soon had a simple transceiver for 80 metres and began to contact some of my old friends on that band.

Inspiration

At about the same time that I was beginning to build my first solid state amateur radio equipment, Doug DeMaw, W1FB, (then W1CER) had recently become the Technical Editor of QST, the journal of the American Radio Relay League (ARRL). He set out to bring the amateur radio constructor into the world of solid-state electronics and published a long series of articles on practical, transistor based, projects. Many of these were simple receivers and low power transmitters.

The Doug DeMaw articles were my training ground in QRP radio construction. Using the QST projects, I began building and using low powered amateur radio equipment. A pursuit, which has held my interest in amateur radio for over 30 years.

Prof. Roger J. Green, M0RJG

My first interest in anything electronics was when my parents bought me a crystal radio set for a Christmas present. In particular, I was amazed at how a radio signal could be received on a bit of wire and drive high impedance headphones, all with no battery – the ultimate QRP radio receiver! This was followed, a few years later on by a kit to build a pair of Bell telephones. Thus started my interest in electronics, and I left junior school to go to the local grammar school and joined an electronics club run by one of the physics masters, Dr. Robin Bellerby, GM3ZYE (now sadly a silent key).

Later, I joined a local radio society, Northern Heights, near Halifax where I learned a lot from other radio amateurs, in particular, Alan Benn, G8AFV. I was so interested in the subject of electronics by then that I went to university to learn more (Manchester, then Bradford) and became professionally qualified in it.

I then realised that I needed to get industrial experience, so I left Yorkshire (a brave move!) and worked for GEC Avionics in Basildon. After two years there, I returned to a temporary post as a lecturer at Bradford University, and, some time later, ended up at Warwick University as a professor. While there, I secretly attended amateur radio qualification classes run at a local college by Michael Dixon, G4GHJ, and managed to get my full ticket!

More recently, in retirement, my main amateur radio interests have become centred on QRP electronics, as low powers and adverse signal-to-noise ratios have been intrinsic to much I have done professionally. I particularly like designing circuits and building them, and, more often than not these days, usually work as designed, as I have had the professional experience of knowing when and to what extent modelling software can be trusted to give the right, real-world results!

Chapter 2

What is QRP?

QRP is low power amateur radio communications. The term QRP comes from the old radio telegrapher's Q code, a system of three letter codes, all beginning with Q, designed to abbreviate standard phrases when sent in Morse code. Sending "QRP" means "reduce power" and sending "QRP?" means "shall I reduce power". So, it is all about reducing the power of radio transmission.

The typical power output of a modern commercial amateur radio transceiver is in the order of 100 watts. The usual definition of QRP operation is 5W of RF (radio frequency) output or less. This is the commonly agreed level for QRP awards and contests and is usually taken as being 5W output for CW (continuous wave) CW (Morse) operation and translated as 10W PEP (peak envelope power) for SSB (single side- band) voice operation, with equivalent power levels for other modes. It is a significant reduction from the usual power levels used in amateur radio communication.

It invites the question: "Why would amateur radio operators want to use such low power levels, and can they expect viable radio communication at these levels?" Perhaps the best

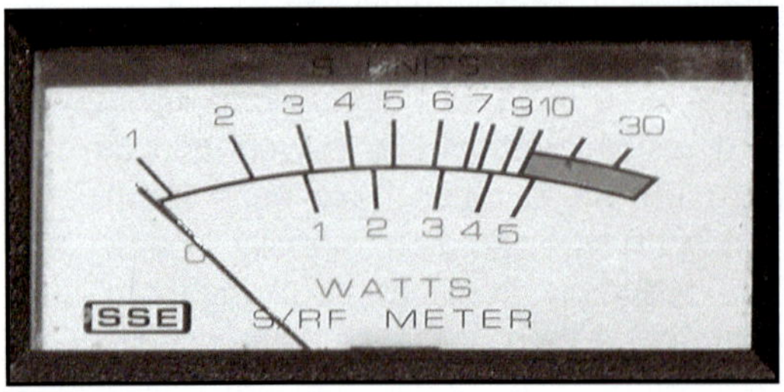

A typical analogue S-meter and RF power meter. Note that both scales are logarithmic rather than linear

TABLE 2.1: SIGNAL STRENGTH	
S1	Faint signals (barely perceptible)
S2	Very weak signals
S3	Weak signals
S4	Fair signals
S5	Fairly good signals
S6	Good signals
S7	Moderately strong signals
S8	Strong signals
S9	Extremely strong signals

way to begin is to consider what power really means in terms of radio communication. The obvious goal is for the receiving station to be able to hear and understand the transmitted signal. Amateur radio uses a signal reporting system called the RST code: Readability, Strength and Tone.

The Readability, using a five-point scale, refers to the quality of the reception, the Strength, using a nine-point scale, indicates the strength of the signal and the Tone, another nine-point scale, describes the note of Morse signals. For our purposes we need only look at the strength. The nine divisions as usually described are shown in **Table 2.1**.

So, for the purposes of reliable communication, S5 signals and above are readable, with S4 and S3 signals being understandable with difficulty. Originally, signal strength reports were subjective. The operator heard the signal and gave a mental assessment of where the signal fitted on the nine-point scale.

Many operators still do that but commercial transceivers are usually fitted with an S-meter calibrated with the nine-point scale. Some S-meters fitted on commercial transceivers are notoriously subjective in their readings depending from where the derive their measurements. But at least the commercial S-meter has introduced a mathematical relationship between the S- units. It is usual for the S-unit divisions on commercial S-meters to represent a 6dB (decibel) change. Armed with this information, let us look at what happens to signal strength as a transmitter reduces power.

The simple chart in **Table 2.2** illustrates what happens as a station reduces

power.(one S-unit is taken to be a 6dB change). The right column shows the power output used by the station. The left column shows the signal strength of the received station in S-units.

The chart begins with the theoretical idea that a station running 1000 watts output is being received at S9. Following down the list, we see what happens as the power is reduced. At 250W, the station is received at S8, at 62.5W it is S7 and is still an S5 signal at 4W. Four watts is well within the usual definition of QRP and S5 is a fully readable signal . . . and that is reducing power from 1000W.

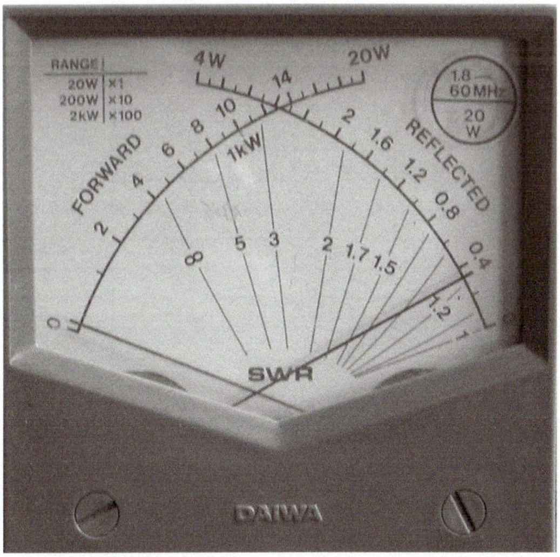

A power and standing wave meter - again note the logarithmic scales

TABLE 2.1: SIGNAL STRENGTH

S-Units	Power output from station
S9	1000 watts
S8	250 watts
S7	62.5 watts
S6	15.6 watts
S5	4 watts
S4	1 watt

The chart assumes that one S-unit change is a 6dB change, which is a four-times power change. The 6dB change per S-unit is because the response of the human ear is more logarithmic than linear, a fact probably forgotten by operators of discos!

So, reducing the power of the transmitter by four times will, in theory, reduce the received strength by one S-unit. Conversely, to gain one S-unit the transmitter power has to be increased by four times. This is a fact of amateur radio life that many operators do not seem to know or just simply do not think about. Using numbers closer to usual amateur radio practice, a transmitter with 100 watts of output can be reduced to 6.25W and will only fall from S9 to S7.

Some people may say that this is meaningless because S-meters on transceivers are often inaccurate. That does not matter. This is a theoretical table and based on the mathematics of the S-meter.

What is more, many QRP operators achieve even better results than this chart suggests. I suspect that is because if an operator is using only small amounts of power, that operator makes sure that as much of the precious power as

possible radiates. Extra care will be taken in the choice of antenna and the matching of the transmitter power to that antenna. Many conventional power operators can be careless with their distribution of RF energy!

The gap between low power operating and higher power operating is not as wide as many people think. With careful antenna matching, that extra skill that QRP operating seems to promote, and the conviction that they will work other stations; QRP operators can achieve quite amazing results. Looking at the logbook of many QRP operators will probably show entries much the same as the logbook of a station using conventional power levels.

An unusual QRP power story

In December 2001, a group of Russian QRP operators mounted a special expedition to the Ai-Petry Plateau in the Crimea mountains in honour of the hundredth anniversary of Marconi's first trans- Atlantic radio transmission. The station, EM100GM, used an old Russian military transceiver type R-143 with a homemade ATU (Antenna Tuning Unit).

The ATU included a dummy load; a 50-ohm resistor for tuning up the transmitter. By accident one of the operators connected the transceiver output to the dummy load rather than the antenna.

Before this was discovered, the transceiver had been in use and made 21 contacts, without being connected to the antenna! How? The dummy load would have been radiating some of the transmitter power to the inductors in the ATU.

The operators suggest that the power going to the antenna may have been in the 5 to 10 milliwatts range. The log book for this very low power period of operation is shown here:

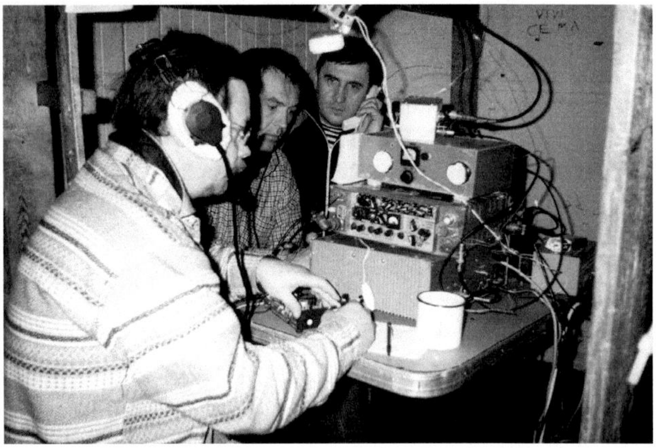

EM100GM in operation (left to right: US1RCH, UU4JCQ and US1REO)

# QSO	Call	Time GMT	Band MHz	RS
1	RK4LXD	14.20	14	59\59
2	UA1UPC	14.21	14	59\59
3	RV9MM/9	14.25	14	59\59
4	UA3BT	14.30	14	59\59
5	RA3DCU	14.31	14	59\58
6	RK4CXK	14.32	14	59\59
7	RW9AS	14.33	14	58\59
8	UA0AX	14.38	14	58\58
9	UA1NCK	14.39	14	59\57
10	UA9FLD	14.45	14	58\59
11	RW4UU	14.49	14	59\59
12	UA4SX	14.56	14	59\55
13	RX3AX	15.00	14	59\59
14	RX3FZ	15.04	14	58\59
15	RK4YWK	15.07	14	58\58
16	RV3DNT	15.12	14	58\58
17	4Z5AF	15.41	14	58\58
18	UR4QI/M	15.53	7	59\56
19	IK4WKU	16.15	18	59\54
20	ON4LAN	16.33	18	59\53
21	PA3DUV/M	16.36	18	59\59

Log entry for EM100GM for contacts made whilst accidentally connected to a dummy load. Estimated radiated power was less than 10mW

Postscript

From time to time I have heard the charge levelled against QRP operators that they are really taking a ride at the other opera- tor's expense. The station at the other end may be using superior equipment, perhaps a superior antenna and certainly superior power. Well there are several answers to that charge. To start with, I am quite happy that others may choose to use higher power or spend more money on their hobby.

Chapter 3

Why QRP?

Although, as we have seen, reducing power is not as drastic as it may seem, why would amateur radio operators choose to reduce their power levels as a matter of course? There are many reasons and motivations and here are just a few of them.

QRP is appropriate

Amateur radio is only a hobby. The word 'amateur' comes from the Latin amare – to love. Radio amateurs do their radio for the love of it. For some this means that spending a four-figure sum on a piece of equipment is not appropriate, or in some cases not possible. In general, QRP stations are low cost stations which allow the excitement of long-distance radio communication without spending a great deal of money. For some people this is their only possible entry point to the hobby. For others it just seems natural to use what will provide interest and achievement without unnecessary complication or expense.

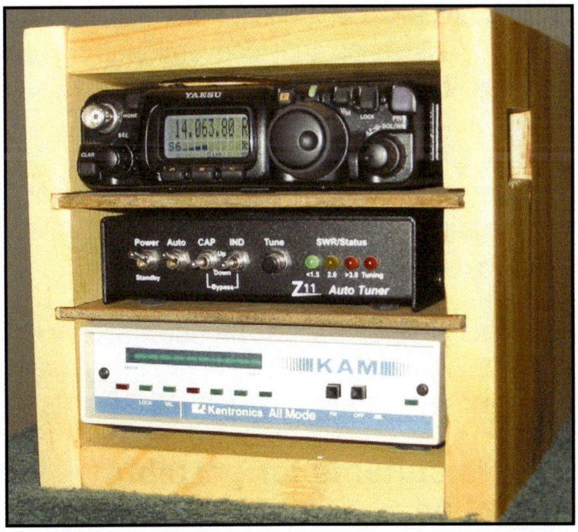

A complete QRP station in a box, based on the Yaesu FT-817 transceiver and built by Anthony Luscre, K8ZT. The transceiver is on the top shelf, a Z11 automatic antenna tuner is in the centre and the lower unit is a DSP filter to aid reception

On the whole, QRP stations are modest in size and will fit into small spaces without the need for a dedicated room or an intrusion into a shared room.

There are many examples of QRP stations fitting into a cupboard or a single small shelf. Some QRPers use the 'shack in a box' approach, a compact station that fits into a wooden or metal box, which can be brought out when required for use. Such a station makes no visual impact on the home until it is in use. Almost anywhere with a small amount of spare space could be a suitable place for a QRP station. QRP can make amateur radio an appropriate hobby for a family home.

QRP is RFI friendly

Many radio amateurs have been plagued by EMC (Electromagnetic Compatibility) problems when the domestic appliances in their own homes or, worse still, in the homes of their neighbours, are affected by their transmissions. The power levels used in QRP operation make such difficulties less likely and often easier to cure when they occur. Some QRO operators have moved to QRP operation because of previous EMC problems and have been able to continue their hobby with more peace of mind.

QRP is a challenge

Amateur radio is a fairly perverse pursuit. Radio amateurs are trying to run a short-wave station from their own homes, capable of making contacts with other stations around the world. Most of them have limited budgets for equipment, only the facilities of an average family home and limited space for the erection of antennas. But this is the challenge of the hobby - doing extraordinary things in ordinary circumstances. The restrictions provide the challenge and overcoming them is part of the pleasure.

Some operators turn to QRP as a greater challenge to their ingenuity and skill. I know several QRP operators who were once successful QRO (high power) operators and found their amateur radio had lost its cutting edge. Reducing power and trying to do it all again has brought a new lease of life to their operating. The satisfaction has returned. QRP operation requires more skill, cunning and subtlety.

There is a whole range of operating awards for the QRP enthusiast and many enjoy their operating in pursuit of these awards. Some awards are particular to individual QRP clubs but recently the ARRL (American Amateur Radio Relay League) introduced a QRP version of their popular DXCC (DX Century Club) award for confirmed contacts with 100 countries using 5W or less. A list of the more popular QRP awards can be found at the end of this book.

D X C C - Q R P Award presented to the late Peter Halpin, PE1MHO, for 100 countries contacted with under five watts on the 6m band

QRP lends itself to home-built equipment

In the early days of amateur radio all of the equipment was home built. Indeed, when I began the hobby, some 40 years ago, most stations were wholly or partly home built. As technology has advanced, and the cost of commercial equipment has fallen relative to salaries, most amateur radio stations are now bought from a shop.

This has turned amateur radio from a hands-on, workshop based, pursuit to a 'user-hobby'. It may be unfair to call it a 'toys for adults' commercially based hobby but the ethos has changed. When I began the hobby, most people understood what was going on inside their equipment - very often they had built it. As more commercial equipment became available at affordable prices, fewer users understood the internal workings of their station. These days it is not uncommon for radio amateurs to buy equipment and never quite understand what all the buttons on the front panel can do.

But there is still a dedicated, and perhaps growing, group of radio amateurs who enjoy building their own equipment. For many of them the QRP option offers viable and affordable stations they can build for themselves. Low power levels mean that conventional circuits and easily available components can be used in their radio construction.

The simplicity of some QRP designs has encouraged amateurs who have never built any radio equipment to begin their first construction projects.

Later in this book a small range of QRP construction projects will be described, most of them suit- able for a beginner and some of them backed up with available kits of parts.

QRP is suitable for portable operation

The relative small size and modest power requirements of a QRP station make portable or 'away from home' operation easy to achieve. Most QRP stations run from 12 volts with a relatively low current drain so reasonable operating times can be had from rechargeable batteries.

An interesting collection of equipment built by Hans Summers, G0UPL

There are several types of QRP portable operator, from those who simply like to take their station out in the car, string up an antenna to a tree, and enjoy operating in the open air to those who go QRP backpacking and operate from remote, mountainous or wilderness areas. Later in this book we will look at the Adventure Radio Club who mix amateur radio and exciting outdoor pursuits.

QRP is a gentle approach to radio

There are some QRP operators who say, that in a wound-up world it is good to have a laid-back hobby. Why use any more power than is needed to have enjoyable radio contacts? Why spend more money than is required to provide a workable station? Why keep buying new radio equipment because last year's model has lost its attraction, or the manufacturers have added another feature which they say you need? Why not build some equipment and enjoy the satisfaction of making radio contacts with something you have built yourself? QRP is simply appropriate.

"Everything should be made as simple as possible, but not simpler"
Albert Einstein

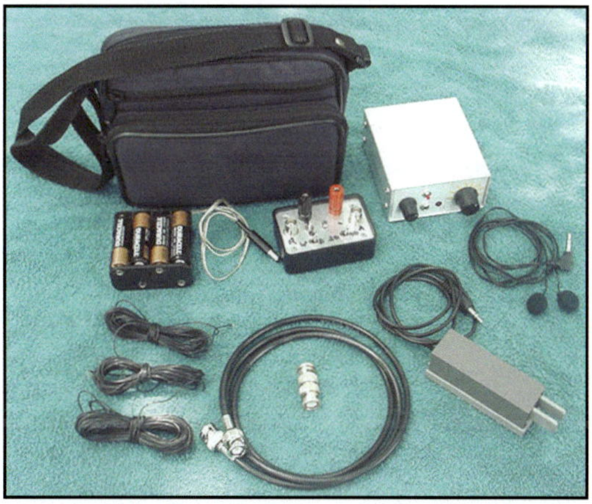

The W0CH 2lb (1kg) radio station: The transceiver used in this set- up is the Wilderness 20m SST; one of the smallest kits available. The SST in the photo- graph has been modified by adding an internal TICK keyer. The SST is used with a simple homebrew tuner / SWR bridge and an AA battery pack. A 33ft (10m) wire is used as a halfwave vertical antenna combined with a couple of 16ft (5m) counterpoise wires. The entire station fits into a small camera bag for easy carrying. Despite this station's diminutive size, it is capable of long range QSOs

Chapter 4

What QRP Operators Say

It may be something about the nature of their chosen interest but QRP operators are a friendly and interesting group. A few well-known QRPers were asked to say something about what makes QRP special for them.

Roger Lapthorn, G3XBM

Roger lives near Cambridge in the UK and has become well-known through his interesting QRP blog at http://g3xbm-qrp.blogspot.com/. The G3XBM QRP Website is at https://sites.goggle.com/site/g3xbmqrp3/home .

His interests include very simple ham radio construction (what you can make for next to nothing) and QRP operating, especially on 6 metres and 10 metres. Roger has done much to promote WSPR (Weak Signal Propagation Reporter) mode, beaconing especially on 137kHz and 500kHz.

"My amateur radio adventure started in 1961 when mum and dad bought me a Heathkit electronics workshop kit for Christmas. Then I was listening to local fishermen on the trawler band and short- wave broadcast stations on an old, and very heavy, DST100 tank radio bought from a local garage for a few pounds. A little later I discovered radio amateurs and I was hooked.

Obtaining a full licence in 1967, my first rig was a simple 400mW QRP AM transmitter for 160m. In the early 1970s 2m beckoned and loads of local stations were worked using a small homemade

10mW AM transmitter and a super-regenerative receiver. Later this was repackaged as the tiny Fredbox handheld powered by a small 9V battery, and one of my greatest thrills was working over 100 miles with this across the English Channel using just a whip antenna one summer morning.

Not long after I discovered 10m and soon realized that a watt or two could get right around the planet with just a small wire dipole antenna when conditions were good. In more recent years HF operating has extended to pedestrian portable SSB with contacts to five continents with just an FT-817 and a small whip antenna and counterpoise.

One notable VHF contact was working a US station by multi-hop sporadic-E on 6m one summer when using just 2.5W ERP. This remains my best DX (distant station) on a VHF band and reinforced a belief that almost anything is possible given a little skill, some good conditions and luck.

There is little to beat the thrill of working DX when running QRP, especially with simple 'kitchen sink' built equipment. An example was in 1984 when an SSB station near Chicago heard my first call, in the SSB part of 15m from

Roger, G3XBM, in his shack

a small homemade 800mW CW transceiver (the Pipit), and a solid con- tact followed. The rig later worked no end of Europeans and DX and became my main rig for several months: I needed no further convincing that simplicity really worked.

My approach remains "keep it simple". Recent designs have been created with minimal test equipment in a small workshop at the back of the garage and I enjoy sharing these ideas with others via my QRP website and blog. Projects don't get finished if too complex, so where possible they are simple ones requiring few parts. In recent years I've started to experiment, simply, at the far ends of the spectrum from VLF to light beam communications.

Despite having enjoyed the hobby for 50+ years now, radio in all its forms still remains truly magical: whether it is receiving a broadcast station from the other side of the world on a tiny crystal set, copying a VLF amateur on 8.97kHz over 1000km away, or being copied across the Atlantic on 10m when running just a few milliwatts.

Joining two wires together with a Morse key and radiating a few milliwatts of radio frequency power that can be detected thousands of miles away across mountains and oceans still sends a tingle down my spine. There is still so much more to explore and enjoy. What a truly wonderful hobby we all share".

Steve Farthing, G0XAR

Steve spent most of his career writing software and managing projects, and when he retired he pottered about in the workshop and travelled the world, as well as his amateur radio interests, including being a member of the GQRP Club and QRPARCI.

"I guess this started for me when I was five, when my grandad Jack, a Northumbrian hill farmer showed me Sputnik 1 flying through the night sky then tuned into its signals on his old valve shortwave receiver. My interest in all things radio was sparked. Then, in the 1960s, we had the space race and the cold war, radio and technology were exciting things.

Steve Farthing, G0XAR

I remember reading the odd copy of Practical Wireless, which in those days had blueprints of projects which all looked exciting. But with very little money, no workshop or tools and no one to 'Elmer' me I had to file it all away mentally and focus on get- ting an education, finding work, building a career and so forth.

Roll the clock forward to the 1990s. I had the career and was driving a desk in the 'corridors of power'. But at the back of my head there was this 'radio itch' which still had to be scratched. By chance I saw an advert in the local paper saying that the Trowbridge club were running classes for the Radio Amateurs Exam, signed up, and the rest, as they say is history. Chris Parnell, G0HFX, taught the course and after a year's study I managed to pass both papers first time. Then there was another year or so learning Morse with Chris. I flunked the first test but passed the second time. In the mean time I had started making my first kits, a Sudden receiver for 20 meters which, considering how simple the design is. worked remarkably well. Then I made a Norcal 40A ready to go on the air with. I was determined to make my first contacts with a station I had built myself! My first con- tact was with Chris, my second with YL2PG in Riga a great distance for 1.4 watts, or so I thought back then.

That was almost 20 years ago yet it seems like yesterday. I have made some great friends

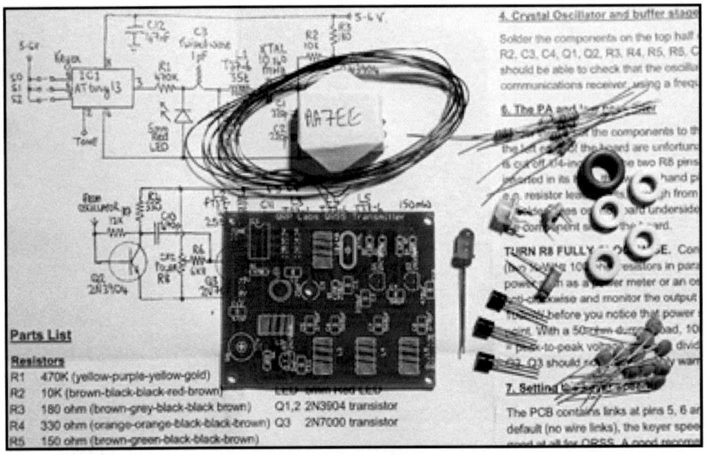

The QRSS Beacon Kit from G0XAR and Hans, G0UPL

through the hobby. Jan G0BBL. Alan M0PUB, Hans G0UPL, Guido PE1NNZ and I collaborate on various QRP projects from time to time. We have travelled the world together and given talks at Pacificon, Dayton and other events. Our designs, notably the QRP2000 and 2001, were at the time at the cut- ting edge of amateur radio.

I have built complex amateur radio projects such as the Elecraft K2 but I still get the most fun out of making simple things out of junk. I have some very nice commercial radios but mostly they gather dust, I only really turn them on when I want to find out what frequency my latest project is working on. On the workbench at the moment is a Twin-T AF oscillator and an audio amplifier which together will make a nice Morse practice oscillator. I rarely operate and my Morse skills need improving so that when I finally get around to building my Sputnik valve Tx I can make some contacts.

Hans and I went into the kitting business a while back and have sold over 500 three-transistor QRSS beacon kits at cost. It is nice to put something back into the hobby. They put out a mighty 100mW on 30 meters which, if conditions are right, can be heard all over the world. QRSS is a derivative of the CW Q-Signal QRS for "Please lower your code speed". By using extremely slow CW, it is possible to use a computer sound card and special software to extract CW characters from below the audible noise floor. Morse code element lengths of 10 to 30 seconds (or even longer) per dot are commonly used. Tiny signals can be heard all over the world.

So, why do I do this QRP stuff? Because ultimately in the grand scheme of things it is pretty point- less. Well I spent all of my career pretty much sitting at a desk being in modern parlance a "knowledge worker". It paid the bills but for me, and I suspect a lot of people but it was not a fulfilling way of life. Nor is watching TV or spending hours surfing the internet. There is something about pottering about with radios that makes me happy. And there is a lot to be said for being happy. Now where did I leave my soldering iron?"

Dominic Baines, M1KTA

Dom Baines is an avid QRP equipment builder and operator. He runs a very active and informative blog at http://m1kta-qrp.blogspot.co.uk/.

"I was licensed in 2003 with a Foundation Licence as M3KTA and then shortly after with a Full Licence as M1KTA. I hold a 5MHz NoV [Notice of Variation to licence conditions] and I was recently granted a 500kHz NoV that also permitted me to operate /P from a couple of locations. The latter is still where almost everything you use has to be built yourself (home brew), although you can use modern equipment for some elements.

Since that date, about 70% of my time spent on the hobby has been spent constructing, both my own projects and those for others less able. I am a somewhat compulsive home-brewer and the shack is pretty well full of radios that might have had one or two test QSOs before I move on to the next project. Amateur radio is a wonderful hobby in which you can build your own receivers and transmitters (once licensed) and there are many opportunities to build your own test equipment and ancillary equipment, antennas and antenna matching units too. Some QRP enthusiasts have built their whole station. The internet has provided the designs of many of the projects I have built as well as complete kits from a number of different vendors.

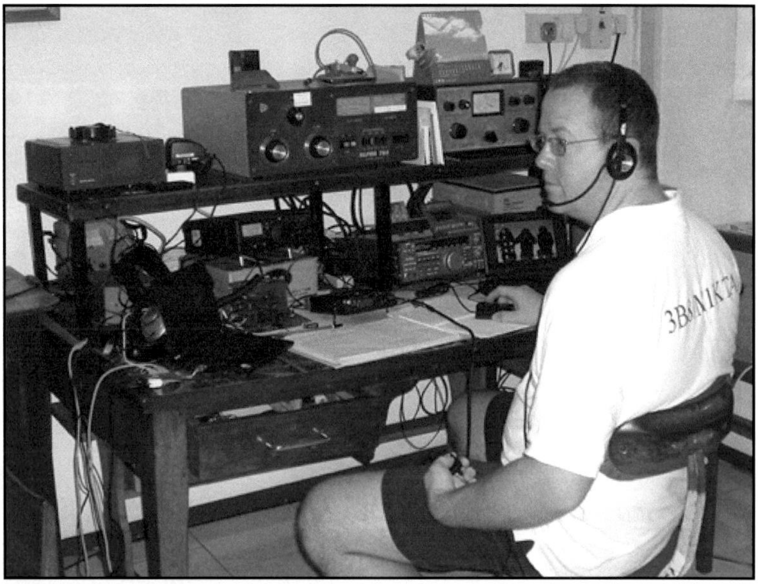

M1KTA operating in Mauritius

When self-building or 'home-brewing', it is often said to be getting harder to find components as rallies are reducing in number. It is true that the older or more obsolete a component, especially some from those classic designs, can be harder to find. However, sites such as eBay and other inter- net traders cannot afford to be overlooked as a potential source of these components by the QRP home-brewer. Think of them as another place to investigate much as you might rummage in a rally stall box. I regularly have parcels of components arrive from all over the globe, both from internet sources or as swaps from other amateurs (do not overlook that option). The net result is that a steady stream of projects is moved one stage closer to completion.

I have no formal training whatsoever in electronics, but this has not stopped me investigating how circuits work. There are many places where you can find information to help you learn. Lately I have revisited topics I originally learnt in order to pass the full licence exam. To that end I bought a NorCal40+ transceiver kit after a copy of The Electronics of Radio by David Rutledge (ISBN 9780521646451) was given to me as a Christmas present. Alongside many other books, it can pro- vide detailed instruction on the theory behind RF circuits.

The remainder of my time has been split between contesting or operating, usually /P from unusual or DX locations. The first occasion I operated in this manner was with another GQRP member, Clive M5CHH, and we put together a mini DXpedition to the Isles of Scilly which lie off Cornwall. The trip coincided with the 2005 RSGB IOTA contest and we operated from a cow shed on St Agnes, the most Westerly inhabited island of the UK. We were joined by another amateur who was holidaying at the time. Whilst not initially planned as QRP we ended up running the station QRP and we were on the air for about 48 hours non-stop, exceeding the original planned contest operating time.

Since then I have operated in many locations and learned that if you have an opportunity to plan wisely, the opportunities multiply for the QRP operator willing to attempt to operate from compromise locations.

In 2007 I lived and worked in Switzerland (HB9/M1KTA) and whilst there I cycled from one end of the country to the other and operated during the trip from batteries recharged using the bicycle dynamo. During this time, I also activated in Liechtenstein (HB0/M1KTA) several times. As many have mentioned that call-sign was "worth an extra kW", and even though these were /P battery powered trips using an FT-817 there were always more stations to contact than I was able to respond to. I worked a lot of US and Australian stations during those trips, and always SSB with just

2.5 watts or less.

In March 2008 I was lucky enough to visit Mauritius (3B8/M1KTA) as my wife's uncle was the science master in a local school. After a rapid set of email exchanges, I secured a licence to operate and a local amateur, Jacky 3B8CF, allowed us to stay with him and I was able to use his 3- and 4- element yagis for 17m and 20m. On that trip I literally worked the world on QRP and had my first experience of a pile up from the other side. I wrote about this in GQRP SPRAT magazine Issue 135.

In 2011, my wife and I cycled 1,000 miles as we travelled the length of the Rhine from near Andermatt in Switzerland to the Hoek van Holland on the coast of the North Sea. I operated in every country we passed through (HB9/HB0/OE/DL/F/PA) almost daily. I was able to operate portable using batteries recharged by the bicycle dynamos. During five weeks operating completely self-contained, I managed a couple of hundred QSOs on CW and SSB. I even had a notebook computer with me, and when the sun shone I managed PSK contacts with solar power.

Whilst in Mauritius I managed to contact 68 different countries despite the propagation not being ideal due to the sunspot cycle and I could not respond to all the stations that called. I have never really understood comments that the bands are closed due to the sunspot cycle as I have learnt that activity usually leads to the bands being lively. Next time there is a contest on, take the opportunity to listen in, even if you do not take part, and you will hear the change in band activity as the propagation changes. Often everything is very quiet in the time just before a contest starts and then again in the minutes after the contest is over. As a QRP operator you can often take advantage of these periods to contact other weak signals, often DX stations. At a gathering of many QRP enthusiasts, the ARCI Four Days in May, in 2008 I listened to a talk by Ward Silver, N0AX, who spoke on QRP contesting advising QRP operators how not to become 'road kill' by thinking about how and when you operate. Ward has a book, Ham Radio for Dummies, which is worth a read and it is worth letting your own family members read if they do not quite understand your fascination with the hobby yet.

My wife has commented several times that if it is possible to include amateur radio during a holiday I often have. One of these days I hope to manage to pull off a QRP trip to a far flung DXCC country, complete with wooden shack on a sandy beach, a dipole strung between two palm trees and a vertical from a fishing pole that does double duty as a real fishing pole, and of course a comfort- able deck chair and rum cocktail in hand for those periods between pile ups.

QRP and amateur radio are whatever you want to make of it".

Bill Meara, N2CQR

In his day job Bill Meara is an American diplomat. He is the author of the book SolderSmoke - Global Adventures in Wireless Electronics, and produces the SolderSmoke podcast and blog (see http://soldersmoke.blogspot.com).

"There I was, out in the Azores islands, in the middle of the Atlantic Ocean, sitting in my radio shack early one morning with a cup of coffee in hand. On the table in front of me was a scruffy looking piece of electronic gear that I had built it myself from bits and pieces of radio junk. It was not much bigger than a shoe-box and it used about as much electricity as my daughter's night light. But that little box had become my magic carpet. In the few short weeks since I'd finished it, it had carried my voice across dark oceans and blazing desserts to the farthest reaches of the earth. With it, I was making new friends all around the world. Not only did that little rig provide the technical means for me to talk to people in exotic far-away places, but the fact that I had built it myself, and that it was using a such ridiculously small amount of power made people sit up and take notice - the nature of the machine itself gave us something to talk about.

In a certain sense, the medium was the message. For me, that device represented the essence of QRP.

Bill Meara, N2CQR

One of Bill's 'Azorean Magic Carpets'

Of course, I could have talked to those distant people with perfect clarity with the cell phone that hung from my belt. But those phone calls would have provided none of the deep satisfaction that I got from the radio contacts made with my ugly little home-made radio. I had created that device; by the time I finished it my hands were calloused and I felt a satisfying sting from the minor cuts and soldering iron burns that you get when you take bits of metal and wire and discarded parts and forge them into a radio transceiver. When I opened the cabinet and looked inside, I saw bits of circuitry that I had improvised, and other bits that distant friends had helped me with (their advice and encouragement coming in via e-mail). As I looked through the circuit, I realized that that radio had taught me things: there was circuitry that I now understood more deeply, and my construction technique - while still far from perfect - had improved. There was a lot of soul in that new machine; cell phones have no soul.

My inclination toward QRP has deep childhood roots. My grandmother had said I was "electrically inclined." My father and I had been big fans of the American writer and radio broadcaster Jean Shepherd. In his nightly broadcasts on a commercial New York radio station "Shep" would sometimes talk about his boyhood experiences as a radio amateur. The great depression was on, so scratch-built homebrewing was the only way he and his friends could get on the air. Necessity was truly the mother of invention. But out of this necessity there emerged a sense of pride, a sense of fraternity and a bit of a technical hierarchy. Status was determined by technical skill and knowledge and by the quality of the rigs they built. I grew up wanting to be like Shep and his friends. I didn't want to be an "appliance operator."

For me, QRP opened the door to that kind of ham radio. Through publications like SPRAT I found a world of simple circuits that I could understand and build myself. I found a welcoming glob- al brotherhood of like-minded radio friends. QRP allowed me to join the technical fraternity described by Jean Shepherd and it led me to the construction of "magic carpets" that put me in direct contact with new friends from beyond the seas".

Mike Rainey, AA1TJ

Mike lives in Roxbury, Vermont. Since his relatively recent return to amateur radio he has become well-known through a whole series of innovative and novel designs for QRP equipment. The 'Bell Ringer' and 'Reggie' became very popular transceiver designs in

Mike Rainey, AA1TJ

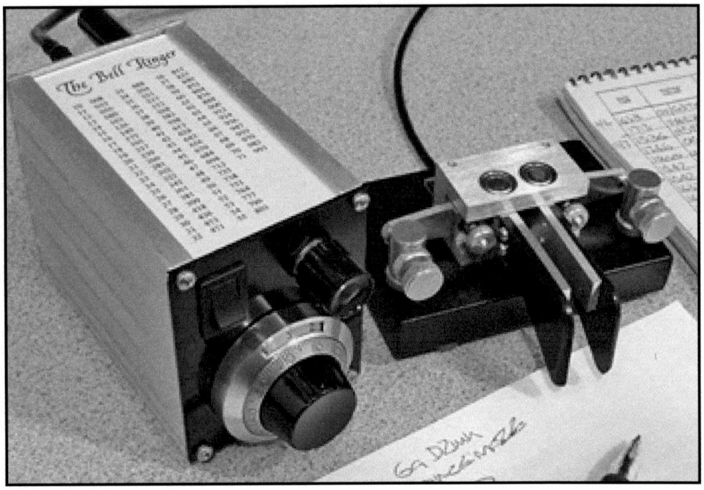

The 'Bell Ringer' transceiver designed by AA1TJ

the GQRP Club magazine SPRAT.

"QRP is bound to an incessant question that may be asked of nearly every facet of the amateur radio hobby. The question is, "Can I do with less?"

In the early days I suppose the question of "less" had mostly to do with how little transmitter power one might successfully employ. Given the astounding historical results with low power transmissions it seems inevitable that hams would increasingly apply the central question of QRP to their entire operation. Can I use less of an antenna? Can I operate my simple equipment from the field under makeshift conditions or using alternative power sources? How many QSOs could I make using a single, non-rechargeable, battery? What is the lowest or highest frequency that I could use to make contacts? What is the least number of discrete components that I could use? Can I build it myself? Can I design it myself? Can I fabricate my own components? What about narrow-bandwidth operation at milliwatt power levels? These are a small subset of the questions that QRPers are actively asking today. Indeed, the challenges are limited only by our imagination.

As for me, I most enjoy working in the realm of milliwatt power levels. I prefer using the simplest equipment, the likes of a heterodyned 'crystal radio' receiver or a tunnel diode transmitter, for example. I take pleasure in using early semiconductor devices. I like prospecting old technical literature for technological dead-ends or otherwise forgotten techniques, with the aim of reproducing them on the work-bench; or better yet, to return them to the air.

A wonderful feature of our hobby is that there appears to be no interest-niche so obscure that you won't find other like-minded enthusiasts. No matter where your own interests lead, rest assured; in QRP you'll meet friends along the way. Amateur radio is inherently a shared experience; our success, and so much of our pleasure, depends crucially upon the operator at the other end. It's little wonder that hams are among the friendliest and technically inquisitive people that you will ever meet. My own special interests are off-beat, to say the least. Nevertheless, by following them I have gained tremendous friendships. Your own interests will guide you to precisely where it is that you belong".

"If you follow your bliss, you put yourself on a kind of track that has been there all the while, waiting for you, and the life that you ought to be living is the one you are living."
Joseph Campbell and the Power of Myth with Bill Moyers.

Chapter 5

What Can Be Done With QRP?

Using low power on the amateur bands does not mean that the operator has to expect under- achievement. There is disadvantage, but as the Power / S-units chart (in chapter 2) showed, the dis- advantage is less than many would imagine. Much of it can be overcome by care, effective use of the available power and operator skill. Most QRP operators try to radiate as much of their precious power as possible by making effective use of their available space for antennas and maximising the

transfer of power to the antenna. They also learn, out of need, the easiest and best ways to secure contacts on the band. They become cunning and shrewd radio operators. In short, they become what all amateur radio operators should be.

GM3OXX(SK)

No account of what the low power operator can achieve would be complete without mention of George Burt, GM3OXX. In the whole of his over 40 years as an amateur radio operator, GM3OXX only used home-made equipment and all of his QRP operating has been done with a power output of no more than 1 watt.

George worked 291 countries on the DXCC list. Sometime QRP operators are accused of hitching a ride on the other stations' power and equipment but GM3OXX especially enjoys working other low power stations. His two-way QRP (both stations using less than 5W) DXCC score is over 100 countries worked, although in his case the power level was under 1W.

Much of his remarkable QRP work was done from Clermiston, near Edinburgh, with a modest wire antenna. The antenna for most of the time was a double extended Zepp for 28MHz with open wire feeders, fed via a balanced tuner unit for all bands. In real terms that was an antenna with a mere 44ft (13.5m) long top. The transceiver, still in use, is a 28MHz CW only transceiver with a crystal- mixing variable frequency oscillator. It tunes from 28.000MHz to 28.200MHz in four switched 50kHz bands. The IF section uses 9MHz, 250Hz bandwidth, crystal filters, so it will only resolve CW signals. The basic transceiver gives a power output of 1mW and is used in conjunction with eight small transverters, one for each of the HF bands, and a small power amplifier for 28MHz. The maximum power output from this combination is 1W.

George Burt, GM3OXX, operates from his former shack in Clermiston, near Edinburgh, Scotland using his homemade one-watt transceiver

STATION WORKED	FREQ	RST OUT	RST IN	PWR OUT	COUNTRY WORKED
08 June 2001					
YO8FR	18.086	599	559	1W	Romania
UX3ZZ	18.086	599	559	1W	Ukraine
UA1QE/QRP [4W]	18.086	589	579	1W	European Russia
HL2FDW	18.081	599	429	1W	Korea
YO4ACC/QRP [3W]	21.06	599	569	1W	Romania
NOUR	21.06	569	449	1W	USA
PP7CI	21.06	599	599	1W	Brazil
8P6BX	21.06	369	569	1W	Barbados
JA2PCT [2W]	21.06	479	559	1W	Japan
JA2DO	21.06	599	599	1W	Japan
OX/LA5KW/QRP [5W]	21.06	579	559	1W	Greenland
VK2GWK	21.014	579	319	1W	Australia
V44KJ	21.02	599	599	1W	Saint Kitts
VP5/N3MT	21.02	599	599	1W	Turks Caicos
JY5HX	14.015	599	599	1W	Jordan
UA9LR	14.06	599	599	1W	Asiatic Russia
09 June 2001					
RA3GD	18.086	589	449	1W	Central Russia
5H3RK	21.012	599	559	1W	Tanzania
EA8RK	21.06	579	539	1W	Canary Is
5Z4FM	24.892	579	539	1W	Kenya
C93MR	24.899	599	449	1W	Mozambique
9G0ARS	24.892	599	599	1W	Ghana
YO4NQ	18.086	599	579	1W	Romania
VP5/N3MT	18.075	599	599	1W	Turks Caicos

The sample extract from the GM3OXX logbook for two days in June of 2001

The current shack of George Burt, GM3OXX, in Netherdale by Turriff. On the bench is the 28MHz basic transceiver with electronic keyer and paddle. The middle shelf holds the band transverters in die-cast boxes. On the top shelf is the balanced antenna tuner with plug-in coils and a standing-wave/power meter

In the sample extract from the GM3OXX logbook for two days in June of 2001 all of the entries are for CW operation using an output power of 1 watt. The entries have been modified to include a list of the countries for those who may not recognise them by their callsign. It is a log that would please any radio amateur, whatever power was being used.

K8ZT

Anthony A Luscre, K8ZT, lives in Ohio and is an avid QRP operator and member of the GQRP Club. He owns a wide variety of QRP equipment and uses various combi- nations of commercial QRP transceivers and ancillary equipment to achieve excellent results. He has several small, self-contained QRO (high power) stations built into wooden units. In chapter 3 we saw his FT-817 station and here is his station based on the TenTec Argonaut 515.

In the extract from K8ZT's logbook for the beginning of 2002 all contacts used 4W output. I have added the countries list on the right - for those readers who may not know all the country callsign prefixes. A glance at this log quickly shows that a good QRP operator can achieve results that are very similar to those achieved using conventional power.

CW or SSB?

G0UKB

Sometimes it is assumed that all QRP operation is done using CW (Morse code – Continuous Wave). CW operation is not every radio amateur's preferred mode; many enjoy operating SSB (Single Side- band) and digital modes. Although much of QRP operating is done using CW, there is a dedicated band of SSB QRP operators and they achieve good results.

Brian Jones, G0UKB, dispels the idea that DX is only possible using CW. Brian is a QRP SSB DXer. He reported to me after working H7DX in a pile-up of stations to bring his QRP SSB DXCC countries count to 150. All contacts were made with a K2 transceiver, with most of them using a 132ft (40m) Windom antenna at 25ft (7.5m). This antenna was lost in the winter winds and has recently been replaced with 102ft (31m) doublet at

The K8ZT Argonaut 515 station

DATE	STATION WORKED	MODE	BAND	OUTPUT (WATTS)	COUNTRY WORKED
01/01/2002	RA6UD	CW	20m	4	Russia
01/01/2002	RN4HN	CW	20m	4	Russia
01/01/2002	7X4AN	CW	10m	4	Algeria
01/01/2002	G3RGD	CW	10m	4	England
01/01/2002	DL6RDR	CW	10m	4	Germany
01/01/2002	HB9KC	CW	10m	4	Switzerland
01/01/2002	ZS6EZ	CW	12m	4	South Africa
01/01/2002	VX9DX	CW	12m	4	Canada
01/01/2002	V51AS	CW	10m	4	Namibia
01/01/2002	GW4GTE	USB	12m	4	Wales
01/01/2002	8P6JA	CW	10m	4	Barbados
01/01/2002	EA4NP	CW	15m	4	Spain
01/01/2002	CM3RW	USB	15m	4	Cuba
01/01/2002	VP9KK	USB	15m	4	Bermuda
01/01/2002	TM4AMD	CW	17m	4	France
01/01/2002	VE1ZF	CW	15m	4	Canada
01/02/2002	5H3RK	CW	20m	4	Tanzania
01/03/2002	HK1RRL	CW	20m	4	Colombia
01/03/2002	NP3SA	CW	20m	4	Puerto Rico
(CW = Morse operation. USB = Upper sideband Telephony)					

The extract from K8ZT's logbook for the beginning of 2002

Anthony A Luscre, K8ZT, in his shack. The operating desk equipment includes (l-r) an Elecraft K2, Yaesu FT- 817, Ten-Tec Argonaut515, Kenwood TS-690 (power turned down to 5W) and computer running LOGic logging Software Anthony lives in Stow, Ohio, USA. Latitude-North 41°12'00.4" Longitude-West 81°24'21.6"

about 32ft (10m) AGL, using a RadioWorks remote balun to allow a short coax feed into the shack for the KAT2 auto-tuner to tune on all bands. A few contacts may also have been on a Cushcraft R7 vertical mounted 5ft (1.5m) above the ground.

Brian writes:

"My first rig was a TenTec Argosy II, and I discovered QRP SSB by forgetting to push in the 50W switch, and contacting an FY5 station with 5 watts by accident. Since then I have been an advocate of low power SSB and simple QRP antennas. Like so many other radio amateurs I have a small gar- den and don't want to upset the neighbours. I am particularly keen to see a raised awareness of QRP SSB in light of the foundation licence. I obtained my K2 kit (#1115) in April 2000 and had it built and enjoyed my first QSO by 30 May. I forget when I reached 100 DXCC score but it was early in 2001. I don't operate too much (I have a day job!) and most of the DX is either during contests or snatched moments, before and after work or late evening. Nearly all the contacts are on 20m and above. My DX includes all continents (including EM1HO for Antarctica) and I have broken into several pileups. I worked D68C on five bands SSB and even three bands CW! I also worked Gerry, VK7GK, in Tasmania on 18m and he dropped his power to match my 10W and we had quite a chat - not bad for 2 stations over 10,600 miles (17,000km) apart. My longest contact has to be ZL2BDW in Wellington, a little under 12,000 miles (19,000km) on the short path. On the QSL card he commented that he hadn't realised I was QRP as my signals were the same strength as the other UK stations he worked.

My most notable DX includes

Brian Jones, G0UKB, operating his K2 QRP transceiver

D2BB, E30NA, FR5D, FS/K4ZA, HH2/F8CUP, HL1CG, VP8CTM, VQ9NL, and more recently H7DX, V51/SP1XF, PW0T, 5U0T and V26S, amongst others. I have also worked 43 US states (including AK. I managed to work 38 states in the 2002 ARRL DX contest.

The station is just the K2 transceiver with the built-in ATU and a standard Kenwood fist mic to the 102ft (31m) doublet at 32ft (10m). I wanted 450-ohm line all the way to an outboard ATU but my wife also uses the antenna from an upstairs room so she wouldn't be happy running downstairs to manually tune the system each time she changes band, therefore I terminate the open-wire feeder externally with a Radio Works 4:1 remote balun, and coax feed into my shack with an option to connect to an upstairs coax line. The K2 will put out up to 15 watts on most bands but I keep it down at 10W PEP which is the accepted SSB QRP level.

A nice example of SSB QRP operation is when I worked Gerry VK7GK (SK) in Hobart, Australia, on 17 metres. He kept telling me I was a weak signal so I kept apologizing for making him work hard. He told me it wasn't hard work he just had to listen a little more and then switched off his linear to drop from 200W to 10W - same as me. He dropped from S5 to S2 on the meter and was still perfect copy - just not as loud. This was my first ever QRP VK contact (and then only my second that broke the 1,000 miles per watt barrier). We had a 10-minute chat and he saved electricity powering his amplifier.

Working with 100W you expect to contact the DX and become disappointed when you fail and with 10 watts you expect not to break the pile-up and get exhilarated when you succeed."

Milli-watting KA1CV, G4BUE

If QRP is a challenge to conventional operating, 'milli-watting' is a challenge to the QRP operator. It is used to describe amateur radio operation at power levels of under one watt. Brice Anderson, W9PNE (SK), was probably the first dedicated milli-watt operator. Brice obtained the Worked All Continents and Worked All States awards and worked some 40 countries using 500mW (half a watt). He used a conventional transmitter with an outboard valve power amplifier. By varying the voltages on the power amplifier valve, Brice was able to produce milli-watt signals.

Ed Hare, KA1CV, ARRL Laboratory Engineer, described milli-watting as "riding the knife edge of what can be done" with amateur radio. His first attempts at ultra-low power operation were with a crystal-controlled transmitter using 74S00 logic chips. The final amplifier was a NAND gate on the chip producing 250mW. In the 1984 CW Sweepstakes contest he worked 24 USA states with his 250mW and by 1987 he had worked all USA states.

Chris Page, G4BUE, enjoyed a long spell of milli-watting at the end of the 1980s. Chris began by using a step-switched resistive attenuator to reduce his signal output. Later he modified a Ten-Tec Argonaut 509 transceiver by reducing the voltage to the power amplifier. With this arrangement Chris achieved Worked All States at 150mW and Worked All Continents at 150mW.

Attempting to see what levels were viable for contacts, in the 1981 CQ WPX contest, Chris worked the following American stations at very low power levels: W9OA (15mW), W0WP (4mW), W1RX (1.5mW), K8HV (625μW), VE3PCA (450μW), KB8SX (200μW) and AB2E (200μW). 200μW is watts!

Carl Mason GW0VSW

Carl Mason GW0VSW at his QRP workstation

Another well-known and keen QRP operator, Carl uses both CW and SSB for his work. He never uses more than 5 Watts on either mode, and uses an Index QRP+ transceiver for most of his activity. In particular, he likes to experiment with antennas for small area locations like his QTH which is not ideal – his home antennas are home brew G5RV, Snowdonia Radio Company vertical, and his own design Loft 'Crown' wire loop. He prefers to do 'more with less' when it comes to QRP. He believes that there is more satisfaction in making a contact with a simple wire or home brew antenna, than a "100-foot-long Yagi"!

His view is that, in amateur radio, communication is a hobby and what is does is not essential for the most part, for example, if a contact is missed because you don't have enough power at the time and/or conditions are poor, there is always another day. He has worked over 100 DXCC entities using both CW and SSB. His best DX to date has been using just 350mW . He enjoys WPX contests and others (all QRP), competing against his own previous scores, depending on the time available. He finds contests help develop the skills which low power operators need, in particular analysing band conditions to judge what's the best path to work a call area at any given time. He enjoys kit building when he has the time. He still holds a couple of GW records with CQ magazine, has been the Chelmsley trophy winner twice, and the GQRP Master #100.

Digital QRP

In recent years QRP operators have become quite excited about the use of digital modes with low power. Most of the excitement has been generated by the ease of setting up and using the PSK31 mode. PSK31 stands for 'Phase Shift Keying, 31 Baud' and is a key- board-to-keyboard communication system using the amateur bands. For low power operation it has several advantages, the chief ones being a narrow bandwidth and good signal to noise ratio. Using the computer screen it is possible to click on a displayed signal and read the typing under conditions that would render its equivalent sound signal inaudible. It digs the signals out of the noise.

PSK31 requires very little hardware and the software is free. The requirements are an SSB transceiver and a basic PC computer with a soundcard. The transceiver has to be interfaced with the computer but the circuit requirements are so simple that many PSK31 users make their own interface connection leads. Several amateur radio kit companies sell kits of parts for making an interface, these products change from time to time but can be found in QRP club magazines and on-line.

The 'panoramic display' software programs (e.g. Dagupan or WinPSK) are free downloads from the internet, and they make it simple to get on the air with PSK31. Readers who want to find out more about PSK31 could go the PSK31 homepage.

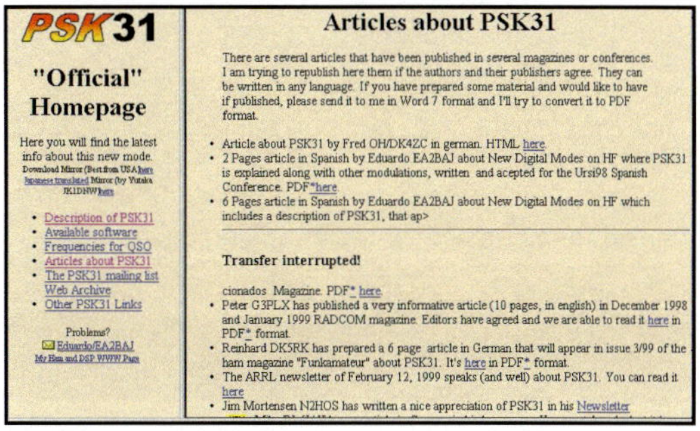

PSK31

"Official" Homepage

Here you will find the latest info about this new mode.
Download Mirror (Best from USA) here
Japanese translated Mirror (by Yutaka JK1DNW) here

- Description of PSK31
- Available software
- Frequencies for QSO
- Articles about PSK31
- The PSK31 mailing list Web Archive
- Other PSK31 Links

Problems?
Eduardo/EA2BAJ
My Ham and DSP WWW Page

Articles about PSK31

There are several articles that have been published in several magazines or conferences. I am trying to republish here them if the authors and their publishers agree. They can be written in any language. If you have prepared some material and would like to have if published, please send it to me in Word 7 format and I'll try to convert it to PDF format.

- Article about PSK31 by Fred OH/DK4ZC in german. HTML here.
- 2 Pages article in Spanish by Eduardo EA2BAJ about New Digital Modes on HF where PSK31 is explained along with other modulations, written and acepted for the Ursi98 Spanish Conference. PDF*here.
- 6 Pages article in Spanish by Eduardo EA2BAJ about New Digital Modes on HF which includes a description of PSK31, that ap>

Transfer interrupted!

cionados Magazine. PDF* here.
- Peter G3PLX has published a very informative article (10 pages, in english) in December 1998 and January 1999 RADCOM magazine. Editors have agreed and we are able to read it here in PDF* format.
- Reinhard DK5RK has prepared a 6 page article in German that will appear in issue 3/99 of the ham magazine "Funkamateur" about PSK31. It's here in PDF* format.
- The ARRL newsletter of February 12, 1999 speaks (and well) about PSK31. You can read it here
- Jim Mortensen N2HOS has written a nice appreciation of PSK31 in his Newsletter

The PSK31 web page:
www.aintel.bi.ehu.es/psk31.html

Chapter 6

Trying Low Power Operation

The best introduction to QRP operation is to try it. There are several ways to get a taste of low power operating with little, or no, outlay of money or special equipment.

The most obvious way is simply to turn down the power of a conventional transceiver. Most commercial HF amateur radio transceivers have an external power control. This may be a knob marked 'Power' or in some cases 'Carrier'. Many commercial transceivers have a nominal output of 100 watts and the power control may reduce the output to as low as 5W, although in some cases around 10W is the minimum available.

There are some problems. At very low drive levels, the efficiency of the power amplifier becomes very poor. In practice, the power being consumed by the transceiver being much higher than the power output does not matter for the purposes of trying QRP on the bands. Low power consumption is an issue for the operator who wishes to use battery or natural power sources but is of little consequence for the casual low power operator at home.

An example of QRP power meter, the Sharman Multicom SWR 12, which also allows measurement of antenna SWR.

How does the operator know what power is being used? Some transceivers have a power output meter but these are notoriously inaccurate at low power levels. They may be LED or LCD based which makes low-level readings difficult or nearly impossible. Again, this really does not matter when trying low power operation. The important thing to is to get a feel for using lower levels of power on the amateur bands. Actual power levels only become an issue when awards are being sought or contests entered. Later in this book I will describe ways to measure power output. A simple way is to buy a 'terminated power meter'. This is in effect a high wattage 50-ohm resistance (the typical impedance for a transmitter output) with a meter to measure the voltage across the resistance produced by a transmitter signal. The scale is calibrated to show that voltage in terms of output power.

If the transceiver has an ALC (automatic level control) input this can be used to reduce the power output. The ALC input is designed to be connected to an ALC control voltage on an external amplifier to control the drive

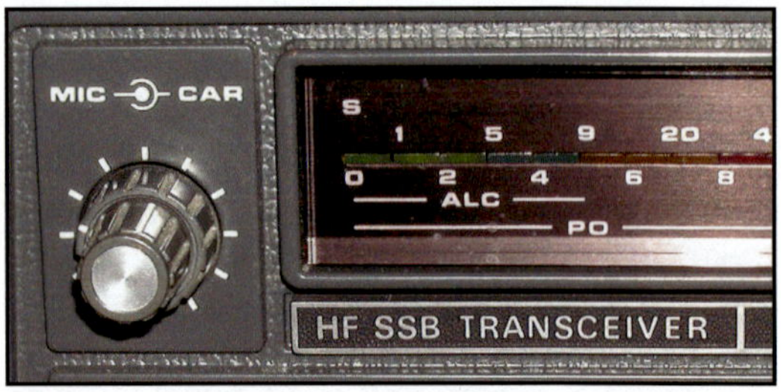

Power control on an HF transceiver (in this case it is marked 'CAR' for carrier)

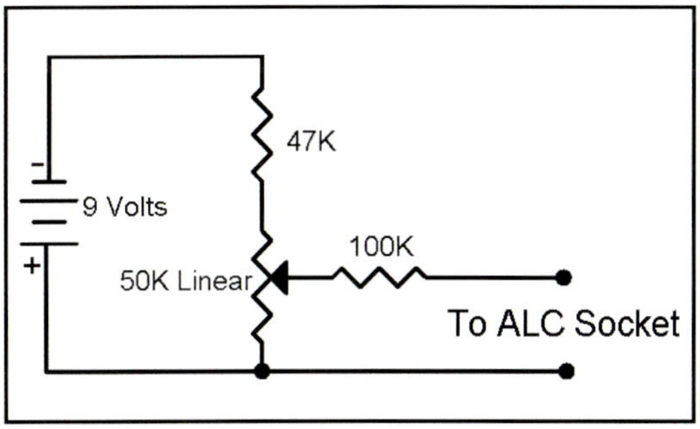

Fig 6.1: ALC Control Circuit

to the amplifier and maintain an undistorted signal in the SSB mode. It is possible to feed a DC voltage to the ALC input and reduce the power output of the transceiver.

Usually this requires a positive voltage, probably in the 4-10-volt range, although this will vary according to the manufacturer's design. A typical circuit for this purpose is shown in **Fig 6.1**.

The manual for a transceiver may indicate the required ALC voltage for power reduction. There are also many simple power reduction circuits and modifications to be found in publications by the GQRP Club and other QRP organisations.

Having reduced the power output of a conventional transceiver, what is the best way to proceed? One idea may be to join other QRP operators. Low power operators tend to gather on, or around, the IARU QRP Calling Frequencies. These are a set of frequencies, agreed by a world-wide QRP group, where QRP operators can seek other low power operators. They are shown in **Table 6.1**. Try calling "CQ QRP" on these frequencies to find other low power stations. They are always pleased to contact other stations using QRP.

A lot of QRP operation is in CW but the speed used by most operators is modest because weak signals may have to be read. If your CW speed is low you should be at home with fellow QRP operators.

Each year the GQRP Club runs an event called the 'Winter Sports'. It runs from Boxing Day (26 December) to New Year's Day and could be described as a QRP 'QSO Party'. It is not a contest or a competitive event but a chance to contact as many other QRP operators as time or opportunity allow. The format is to call, or listen for, "CQ QRP" on, or around, the International QRP Frequencies and simply enjoy the contacts. GQRP Club members usually exchange their club membership numbers and there is an award available, the G4DQP Trophy, for the best overall contribution to the event. But anyone is welcome to join in the event and many operators have had their first taste of QRP by turning down the power and joining in the Winter Sports.

If I were trying QRP for the first time, I suspect I would try some CW QSOs on the 20m (14MHz) band. It is a good general band for low power operation. In Europe a fair range of countries can be contacted using a few watts and a modest antenna. In the 1980s I used to operate portable from a cottage in Wales using 2W output to a 20m dipole strung between the cottage eaves and a low tree. I used to set myself the goal of working 25 countries in a week, and achieved it several times in casual evening operating.

20 metres is also a good band for finding stations which only want short 'rubber stamp' type contacts; an exchange of names, loca-

TABLE 6.1: INTERNATIONAL QRP CALLING FREQUENCIES

CW:	SSB:
1843	3690
3560	7090
7030	14285
10116	18150
14060	21285
18096	24950
21060	28360
24906	
28060	

Simple low power transmitter on a cardboard box

tions and basic station information. These are ideal for trying low power contacts. The information exchanged is simple, repeated several times, and easy to send and read. It is better to answer "CQ" calls than call "CQ" and it can be fruitful to 'tail-end' another QSO. That is waiting until an existing contact draws to an end and call one of the participating stations. Simple QRP operating techniques will be dealt with later on in this book.

Using a simple low power transmitter with an existing receiver, or the receiver portion of a transceiver, is another way to taste QRP operation. The are many designs and kits for building simple low power transmitters. It requires little skill or money to produce a couple of watts on an amateur band (see later for practical designs and applications).

Many QRP operators began their low power amateur radio in this way. A small transmitter board alongside a receiver is a fun way to operate. For many people it has provided a first experience of making contacts using equipment they have built themselves. But be warned! Going on to the amateur

bands and having contacts using a little home-made transmitter can become addictive. Operators trying this for the first time have become so pleased with themselves that their expensive commercial equipment gets pushed aside and they become hooked on using their little home-made creations.

Do not assume that QRP operators are like some esoteric order, the amateur radio equivalent of Trappist monks. There are plenty of dedicated QRP operators who only use low power but most QRP signals come from stations which also include conventional equipment but use QRP to add a little variety, or spice, to the hobby. Most members of QRP organisations are ordinary radio amateurs, with some sort of conventional 'plumbed-in' station who also run a QRP transceiver or who enjoy building QRP projects and using them on the bands. QRP operation is a cheap and simple way to expand the horizons of an individual's amateur radio.

Chapter 7

What is a QRP Station?

There are as many types of QRP station as there are QRP stations. That is part of the fascination. Go into any conventional amateur radio station and it is often much the same as the next - the only thing that varies is the make of the equipment. Go into a QRP operator's station and it will probably reflect the individuality of the owner.

Some QRP stations are neat and compact to meet the needs of space or domestic convenience. Some QRP stations clearly show the practical or experimental nature of the owner's hobby; a col- lection of complete and partly built equipment. Some are examples of the hobby of amateur radio at its best and others, like mine, are simply a mess! Let us take a look inside some QRP stations.

The neat and compact QRP station is a popular choice for many operators. It allows a hobby to integrate with home life and is easily concealed and transported.

The operating corner of the G3RJV 'mess'

One small area in the corner of a room can house a complete station as can be seen in this home station of Dave Aldridge, G3VGR

Another example of a K8ZT 'station in a box': an Argonaut 509 station with matching Ten Tec accessories

The photograph above shows the QRP station set-up at G3VGR's home location. An MFJ digital clock sits on top of a Wilderness Radio Sierra transceiver. This is a kit-built transceiver, which uses plug-in modules to choose a range of amateur bands. A Bencher BY1 paddle is to the side. On the shelf above are a Timewave DSP9+ audio filter, an LDG automatic QRP antenna tuner and some plug-in band modules. The LDG tuner has been modified to include an internal balun for balanced antenna feed lines.

Tony Fishpool, G4WIF, in his QRP workshop combined with his station

As QRP operation is often closely linked to the building of equipment, some QRP stations are practical workshops. That station may be located within the workshop and construction and operating become integrated activities. Tony Fishpool, G3WIF, is an avid QRPer. He is a keen builder of equipment with several published articles on test equipment and maintains the GQRP Club website. He is the joint author, with G3MFJ, on a book of test equipment for QRP constructors.

Expect to find a variety of equipment in many QRP stations. Home built equipment can be delightfully diverse. Hans Summers, G0UPL, has built and used many types of QRP equipment. Although most of his home-made projects use recent technology, he has built, and used, items using valves.

There are many kits available for the QRP station and I will review a variety of these later in this book. Building a kit, or kits, is a convenient and inexpensive way to set up a QRP station. Mark Coultas, G0SLP, used Howes kits (England (~1980`s - 1990`s) to make up a smart functional station.

QRP stations in eastern Europe have always impressed me. Radio amateurs in this part of Europe are emerging from the period

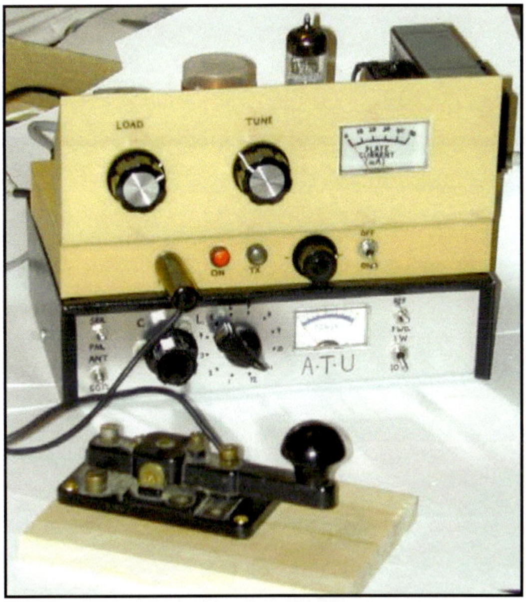

A one valve transmitter for the 80m band built by Hans Summers, G0UPL, based on a design by Jan Axing, SM5GNN. It uses a single Siemens ECL82 triode / pentode valve (US equivalent 6BM8) to produce about 8W. The transmitter sits on top of a homemade ATU and uses an ex-RAF straight Morse key

The kit-built station of Mark Coultas, G0SLP. It consists of a Howes 21/28MHz SSB/CW transceiver, ATU, audio filter, plus a kit-built frequency counter. A neat two band, two mode, station

when they found commercial amateur radio equipment almost impossible to find. But they bring with them the ingenuity and home construction skills that have made their stations interesting and innovative. In my role as the editor of the GQRP Club journal, SPRAT, some of the most creative circuits and designs I have received have come from eastern Europe.

Gary Podgorny, EW1MM, in Minsk, has always used home-made amateur radio equipment. Gary's station is a mixture of equipment built up over the years and includes one amateur radio commercial transceiver, a Ten Tec Argonaut transceiver from the 1970s. The main QRP transceiver is an innovative modification of a surplus communications receiver. Seeking a frequency stable basis for a station running digital modes, Gary obtained a Russian communications receiver type R160P. He installed an encoder and converted the receiver into a QRP transceiver.

There are no accepted patterns for setting up a QRP station. It depends upon aims, means, space and commercial and homemade alternatives. Later we will look at a range of commercial, kit-built and home constructed equipment that could be used for low power communication.

The varied station of Gary Podgorny, EW1MM. The large unit at the end of the table is the R160P receiver converted to function as a transceiver

Chapter 8

QRP Portable

One of my 'radio pleasures' over the years has been using a compact QRP station in a portable or holiday location. QRP lends itself to operating away from home. For some this is merely packing a small station to use in a hotel room. For others it is backpacking, or even operating from mountain summits, with a completely portable HF station. Many QRP transceivers, almost by definition, are small and easy to transport, the power requirements are modest and there are many compact HF antennas to choose from. Some operators of conventionally powered home stations turn to QRP operation to add a portable dimension to their hobby. It is simply a good way to enjoy amateur radio. Travelling away from home can be made more enjoyable when it is combined with a little amateur radio operation. Dave Aldridge, G3VGR, uses his Sierra transceiver from hotel rooms with a homemade power supply and homemade iambic paddle. Dave uses a half-wave end-fed wire with counterpoises for each band.

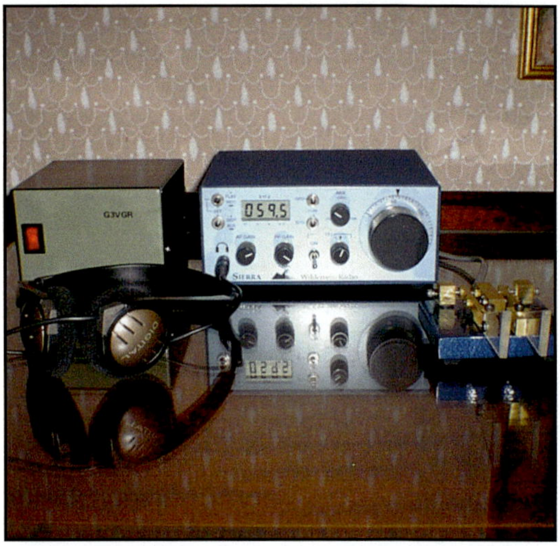

The neat G3VGR portable hotel station: (l to r) Homemade power supply, Sierra transceiver, home- made paddle

The G3VGR end-fed half wave antenna for portable locations is based on an idea from M0AJL which appeared in the GQRP Club journal SPRAT. A parallel tuned circuit (**Fig 8.1**) matches the half wave wire to the 50-ohm output of the transceiver. G3VGR used toroidal cores for his tuned circuit.

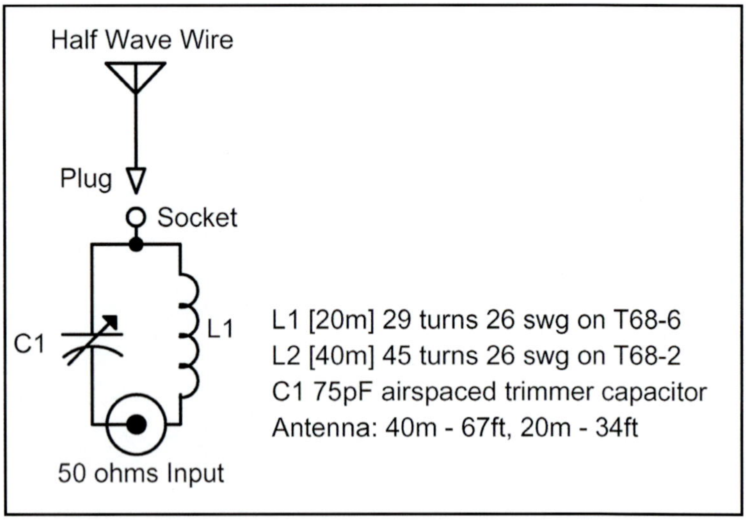

Fig 8.1: G3VGR uses a parallel tuned circuit to match his half-wave wire

The great outdoors

Avid QRP portable operator, Adam Nathanson, N4EKV, lives near the Briones Regional Park in Contra Costa County, California. There is a beautiful hilltop in the park at about 1200ft (370m) where Adam sets up his 1W 20m CW station. The views are glorious and a park bench provides a sturdy support for Adam and his antenna. The transceiver is the kit-built Small Wonder Labs SW-20 with a Palm Mini Paddle attached to the right side of the cabinet, powered by a 12V 2AH sealed lead acid battery. The SW-20 has a VXO range from 14.025 to 14.070MHz. Adam's antenna is a DK9SQ (SK) fibreglass telescoping mast supporting a 20m ground plane made from wire. The basic DK9SQ pole is 10 meters tall, but Adam removed the top two sections to give a total height of about 25ft (8m).

Adam's operation is a prime example of the pleasures of low power amateur radio. An inexpensive small transceiver, that has the added satisfaction of being made by the operator, linked to a day out in glorious open-air scenery and the bonus of contacting other stations with the minimum of power. It is not surprising that portable operation has drawn so many radio amateurs to QRP.

N4EKV enjoying QRP portable in Briones Regional Park, California. The close-up shows the small Wonder-Labs SW-20 and Palm Mini-paddle

QRP from a bike

Andy Meng, N8MX, is from the Finneytown area of Cincinnati. A keen QRP CW operator and cyclist, Andy decided to combine his amateur radio and cycling. His initial approach was to mount his Elecraft K2 transceiver on his bike by just hose clamping the K2 tilt bail bar to the cycle handlebars. The antenna was a full quarter-wavelength 10m whip mounted on a bracket on the back of the bike. The arrangement worked but looked ugly so Andy tried a more radical approach.

He decided to remote-mount the front panel of the K2 on the handle bar of the cycle and connect it by ribbon cable to the main part of the transceiver. The main transceiver electronics were housed in a custom-made aluminium case under the cycle cross bar. The set-up was completed with helmet-mounted speakers and microphone and a NiMh battery pack. Although the N8MX K2 transceiver has the SSB option, Andy also wanted to try bike mobile CW operation. So, a set of 'Poor Man's Paddles' (a simple paddle key arrangement made from printed circuit board material from a QST magazine article by N9JXY) were mounted beneath the handlebars. The whole arrangement allows CW or SSB operation 'on the move' and is a fine example of the ingenuity of one young radio amateur.

The more usual format for bike mobile operation is to use the cycle to transport the station to a chosen location and then operate as a static portable station. This is the easy way to combine some exercise, fresh air and amateur radio.

Steve Rawlings, GW4ALG, is a keen cyclist and when operating QRP, used this approach. On hearing that I intended to include a section on portable operation in this book, Steve was kind enough to send me a lot of information on his approach to cycling portable operation.

Here are some of Steve's own words:

"Until a few years ago, most portable operation involving hand-carried (or back-packed) equip- mint was limited to carrying a VHF/UHF 'walkie-talkie' to a local hill-top. But the growing interest in constructing effective and compact HF QRP rigs – both scratch-built, and from kits – plus the availability of commercial HF 'handheld' equipment, means that HF back-packing has become a popular way for amateur radio operators to spend a day in the countryside. Indeed, it is this ready availability of portable equipment (as well as light-weight fibre-glass antenna supports) that means

it's never been easier for those who enjoy 'adventure radio', to seek out interesting places from which to operate their portable radio equipment."

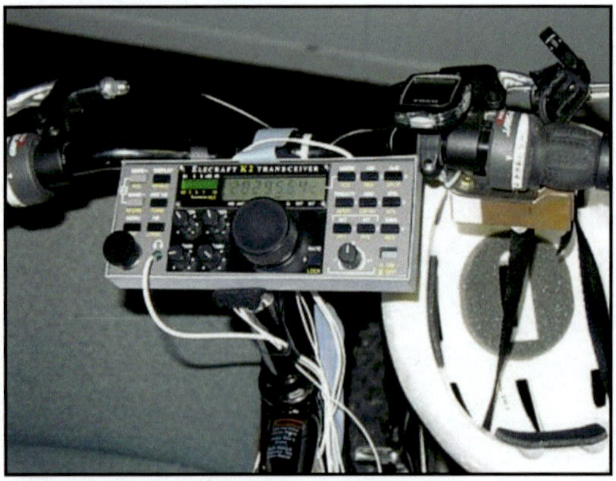

Andy Meng, N8MX, mounts his transceiver below the crossbar with a remote front panel on the handlebars

The GW4ALG bike portable station

Why 'bicycle portable'?

Many of those who go 'back-packing' usually need to drive for at least part of the way to their destination. This provides a quick method for getting to and from the site, and also enables rapid transit to an alternative site, should this be necessary.

In my case, there are plenty of potential sites within cycling distance of my home in Chepstow, south-east Wales. A comfortable cycling range for me is about 15km (10 miles). But even this limited radial distance takes in large stretches of the beautiful Wye Valley; the local high point at Wentwood Hill; the unique Gwent Levels; and the scenic villages and open countryside to the north of Bristol. But I don't think that my location is particularly unique - I suspect that most UK amateurs can think of several potential portable locations within a 15 km radius of home.

There is a certain sense of achievement to be gained when transporting all the essential items of radio equipment; provisions; and safety equipment by 'pedal power' alone. Apart from anything else, the more leisurely pace of touring by bike enables a much better appreciation of what's around you than other forms of transport.

Of course, cycling is a great way of keeping fit, too! I am advised that, for most people, cycling is kinder on the knees and hips than other form of exercise, such as road running. Of course, if you're in any doubt about your own health, it would be worth consulting your doctor before commencing any form of strenuous activity, such as cycling.

What bands?

Most bicycle portable activity takes place during the day, so you'll need to prepare for operation on bands that are likely to provide a reasonable chance of making some contacts during the hours of daylight. For example, the lack of daytime ionospheric propagation on the lower frequencies may make 80 metres a poor choice. Even though 160 metres suffers from the same problem, a special operation using, say, a kite-supported vertical wire, might generate enough local interest to make the operation worthwhile – especially if the outing is announced beforehand at the local radio club.

What antennas?

For VHF/UHF, a collapsible Yagi or similar antenna is a good choice, together with a short mast. Sometimes, such masts can be supported by a fence post, or a sturdy sign post.

But most of my trips have involved using the bike frame itself as a support. After first guying the bike in an upright position using two short guy ropes from the seat tube, I then tie an extended telescopic mast to both the bottom bracket and the crossbar of the bike. A couple of layers of 'bubble-wrap' around the bike tubes helps to protect the paintwork from damage; and the added friction introduced by the bubble-wrap helps to keep the mast firmly in position.

For LF/MF/HF operation, the obvious choice is a wire antenna in the form of either a vertical; end fed wire; or centre-fed doublet. After carrying out early experiments using vertical and end-fed antennas, I now prefer to use balanced antennas for 10 – 80m because they don't rely on a good RF earth for effective, predictable results. A resonant dipole, 50-ohm twin feeder, and balun work well together. However, my favourite antenna (for both single-band and multi-band expeditions) consists of an inverted-V centre-fed doublet; a balanced feeder (using lightweight speaker cable); a switch- able 1:1/4:1 balun; and an antenna matching unit.

Most lightweight telescopic poles and short masts can be transported by being tied to the cross- bar of the bike. Although all the equipment and batteries can be carried in a back-pack, this is far from ideal. The better, and much safer, option is to fit a rear rack to the bike, and use panniers for the heavy items. If I still need to take a back-pack, I use it for lighter items only - such as waterproof clothing.

Other antenna options worth trying on the lower frequencies include kite- supported antennas. For an RF earth, you can try using crocodile clips to con- net the ATU earth-point to nearby fence wires. Note that running out your own ground wires can take a surprisingly long time, and such wires may present a safety hazard to you and to passers-by. Also, keep in mind the safety considerations when using kite antennas. Apart from the obvious need to ensure that such antennas do not come into contact with other structures/power lines, it is also important to ensure that the antenna does not break free, and add to the litter, which already spoils our countryside and coastlines. Please remember: what goes up, eventually comes down.

TABLE 8.1: PACKING LIST FOR BIKE MOBILE

- Two litres water;
- Lightweight waterproof jacket and leggings;
- Basic first aid kit (plasters; bandages; etc);
- Sandwiches; fruit; breakfast bars;
- One (or two) spare fleece, sweater, or similar extra 'layer'; gloves;
- Tyre pump; spare inner tube; puncture repair kit;
- Tool kit, comprising tyre levers; Allen keys, multi-purpose knife ('Swiss Army' knife, or similar) and spanners;
- Small groundsheet, or large carrier bag;
- Guy ropes, guy stakes, hammer;
- Transmitter / receiver;
- Battery;
- Connecting leads;
- Antenna and connectors;
- Mast / pole assembly;
- Cable ties; reel of insulating tape; a few lengths of string;
- Morse key, microphone;
- Paper, pens, watch, clip-board;
- Map, location information (SOTA/WAB/ Maidenhead Locator/county etc.)

In all cases you should aim to leave your chosen location 'as found'. Ensure that all odd pieces of tape; string; wire; and other rubbish are securely stowed away for the return journey. Leave only your footprints and tyre marks behind.

What to pack?

It's all too easy to pack the rig and an antenna, hop on the bike, and take to the road - and then end up being ill-prepared for a change in weather conditions, or perhaps a sudden change in tyre pressure! **Table 8.1** lists in no particular order, a suggested 'starter' packing list for your first bicycle portable outing. Other added luxuries might include: a fold-up stool / chair; compass; GPS receiver.

Touring is almost always safer (and much more pleasurable) if you have a companion. If you can't persuade someone to join you on your adventure, you might wish to take along a mobile phone in case of emergency.

Where to operate?

Unless you already have a specific site in mind, you will probably need to study a map of the area to come up with some possible locations. But determining the exact site will require allowing time to carry out a 'survey' of the general area, taking a number of factors into account. Not least will be considering how best to support the antenna, and finding a comfortable operating position.

To operate for more than a few tens of minutes, establishing a comfortable operating position is very important - especially when operating CW. On my very first outing, I was lucky enough to find some large logs to sit on. After guying the bike in an upright position near one of the logs, I then fit- ted my home-made plinth to the top of the rear rack (to act as the 'operating desk'). The Elecraft K2 transceiver, ATU, SWR bridge and Morse key were then mounted on the plinth. The resulting sitting position was so comfortable that, during the many fine contacts made that day, the old expression 'armchair copy' really did seem to be quite appropriate! From other locations,

I have also used tree stumps, and upturned plastic buckets as seats. For me, using almost any kind of seat, together with an elevated Morse key is far more comfortable than trying to operate CW with everything at ground level.

When choosing a site, I look for a spot without road noise, and well away from power lines. On cool days, try to select a site having some shelter from the wind; on very sunny days, a shady spot will reduce the risk of sun burn, and make it easier to see the front panel displays. Be sure to ask per- mission from the landowner before setting up your station on private property.

Results

Operation from almost any location will produce results that are at least as good as the results that you would expect from your fixed station at home. OK - perhaps your antenna will not be as high as the antenna at home – but it is likely to be well away from other conductors (such as house wiring), and sources of electrical noise. Also, you will find that your novel portable set-up will be of great interest to the stations contacted. I have often received the comment: MY FIRST QSO WID BIKE PORTABLE STATION HI!

The results while operating portable will be limited only by the prevailing propagation conditions, and by your own ingenuity. There is much scope for improving antennas and equipment, and for refining techniques for stowing the equipment on the bike; setting up the portable station; and establishing a comfortable operating position. Above all, have fun - and enjoy the countryside.

Adventure radio

Adventure radio is an idea gaining popularity amongst many radio amateurs who enjoy outdoor pursuits. It is the other side of the *"let's take our QRP station into the open air"* coin. True adventure radio puts the adventure first and the radio second. An adventure radio website defines it as: *"the use of radio to complement adventurous human powered travel"*. Wes Hayward, W7ZOI, sums the adventure radio philosophy in his classic book Solid State Design for the Radio Amateur by saying,

". . . the radio amateur who carries his hobby into the back country should be prepared for the occasional emotional dilemma. Should he compromise his hiking or climbing goal in order to get on the air or should he pursue the primary goal? In this age dominated by high technology, the answer is obvious. Climb the mountain!"

Just as the radio is secondary to the adventure, so is the QRP secondary to the radio. The adventurer probably wants to use as much power as is physically possible but the constraints of weight, size and power supply source will usually result in a station using 5W and often less. Thus, many radio adventurers build special radios that are easy to use in adverse conditions, or modify existing kits to suit their needs.

A good starting place to find out more about adventure radio is the Adventure Radio Society. In the UK, Richard Newstead, G3CWI, is a

Richard Newstead, G3CWI, operating portable on 40 metres CW using an Elecraft K1 at the summit of Caer Caradoc hill in Shropshire. This operation was on the first day of the Summits on the Air programme of which he is a founding member.

Sat 20 Jan	
08:55	Set off from Pen-y-pass in near-perfect conditions: calm and sunny. Ascended via Pyg track.
11:35	Reached Snowdon summit. Set up aerial behind Mountain Railway Station
11:57	First contact made. Fourteen contacts made within the allotted hour.
13:30	Departed from Snowdon summit.
15:35	Arrived back at Pen-y-pass
Sun 21 Jan	
09:50	Set off from N. end of Wast Water to ascend Scafell Pike via Brown Tongue Gill and Lingmell Col. Poor conditions.
12:50	Reached summit of Scafell Pike. Conditions too poor on summit for transmission.
13:30	Stopped just below Lingmell Col and set up aerial.
13:56	First contact made.
14:16	Transmission stopped due to cold and poor conditions. Six contacts made.
14:30	Set off back
16:00	Arrived back at Wast Water.

The above description of the Mint Box Challenge shows what an adventurous radio amateur can do with simple equipment spurred on by a charitable intention

prime mover in the adventure radio movement and his web- site at http://www.qsl.net/g3cwi/ contains a lot of interesting and useful information. His new YouTube channel shows even more: https://www.youtube.com/user/g3cwi . Richard has also been instrumental in setting up the Summits On The Air (SOTA) programme (http://www.sota.org.uk/) for those who enjoy high altitude adventure radio.

The Mint Box Challenge - G3CWI on the summit of Snowdon

The Mint Box Challenge

One of G3CWI's adventure radio expeditions which captured the imagination was the Mint Box Challenge. Shortly after Kanga Products (https://www.kanga-products.co.uk/) introduced their little FOXX-3 transceiver kit, which fits in an Altoids mint tin, Richard decided to use a FOXX-3 for charitable fund raising. Richard's local church supports two young missionaries working in Siberia. The plan was to climb Snowdon and Scafell Pike (the tallest mountains in Wales and England) in a weekend, to operate from each one for an hour and be sponsored for each contact that was made. The results make interesting reading:

An unusual portable story

Portable operation often involves improvising. The operator is away from home and without a work- shop or sources of spare and replacement parts. Tim Allison, GOTYM, tells an interesting story about portable operating

The 'Miracle Whip'

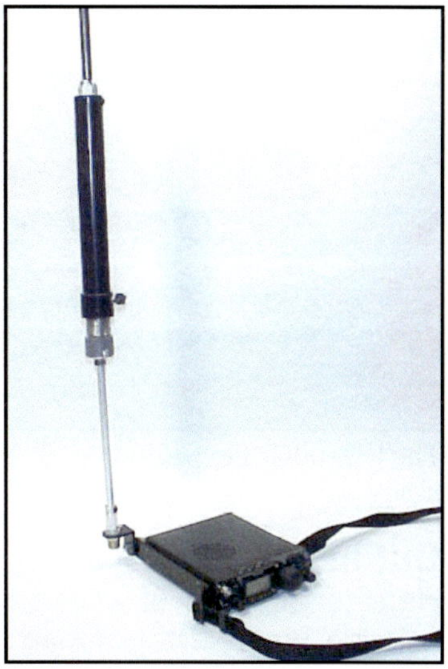

The Super Antennas "MP1" mounted on an FT817

in Germany. Tim was attending the Friedrichshafen *HamRadio* Hamfest and had taken his Mizuho MX14s handheld 14MHz CW and SSB transceiver. He took a break in a small chalet style bar, a popular meeting place for radio amateurs, on the top of the Pfander mountain in nearby Austria. The mountain is 1064m (3500ft) ASL, and perched on the top is a large communications mast, which also houses the local amateur repeaters. This seemed an ideal place to try his portable station, then he discovered he had left his Morse key in the hotel room.

Paul says: *"I discovered my Watson key had been left in the hotel room. However fellow traveller Ernie, G4LUE, came to the rescue with his improvised key. As I was struggling to get signals out using the two ends of a spare power lead taped to two screwdrivers, Ernie spotted my Pentel pencil laying alongside the logbook. In a flash he had the spring clip moved and had trapped the leads beneath it and the metal pencil end. Now we had a key which could be operated by holding it in one hand and keying using the index finger on the spring clip. The TY KEy was born. 'Proof of the pudding' was a first con- tact into Devon with G0GMA/P."*

Equipment, power and information

QRP portable operation is a delightful way to enjoy amateur radio. There are many choices for building up a light-weight, low power demand, station - many of them appear in this book. The advent of the Yaesu FT-817 transceiver encouraged many people into portable operation. The selection of a suitable antenna can be a challenge. Some operators make use of fibreglass telescopic poles (see traders appendix) to support simple wire antennas. There are several commercial portable antenna options. I have had good results from the MP1, designed by Vern Wright, W6MMA(SK), of Super Antennas (see traders appendix).

Other QRP operators have praised the Miracle Whip antenna, a small loaded whip with a built-in antenna tuner. I am not sure about the word 'miracle' but it certainly does what can be asked of a loaded whip antenna and is very convenient to use . It is purpose-made to fit the FT-817 transceiver.

Power supplies need to be a compromise between size/weight and capacity. Lead-acid GellCell batteries are commonly used but the higher capacity nickel metal hydride (NiMH) batteries, designed for digital camera and video use, or modern Lithium batteries are attractive alternatives. As with all re-chargeable batteries, trickle charging is a safe method, and if rapid charging is being done it is best with a charger that monitors, or prevents over charging. Some more 'eco-aware' operators use solar cell charging for their batteries.

Those wishing to become frequent and more adventurous portable operators may like to join one of the Internet based radio portable groups. The Adventure Radio Society, Summits on the Air, HEMA and the HFpack, portable HF operating group, can all be found in the QRP Organisations section at the back of this book.

The EN100GM expedition

G4AKC with his solar-powered, trolley-mounted station (more pictures at http://www.qrz.com/db/g4akc)

Out in the cold

Here is mention of a very intrepid QRP portable group. The Russian UR-QRP Club mounted an expedition in December of 2001 to celebrate the 100th anniversary of the first Marconi transatlantic radio contact. The expedition was to the Ai-Petri plateau in the Crimean Mountains. Expedition mem- beers were Dmitriy, UU4JCQ, Igor, RK3ZK, Peter, US1REO and Victor, US1RCH. Using the callsign, EN100GM, they made over 1,000 QSOs in a single week in very cold and difficult conditions. The photograph shows the group marking the Marconi anniversary day on 12 December by operating a 1W CW transmitter using a dipole antenna attached to a tree. RK3ZK commented that the equipment worked well; better than the operators. The chief problem was using a Morse key with numb hands!

G to ZL using 2W SSB solar-power pedestrian mobile

Most QRP operation uses CW and sometimes I am asked if low powered operation is viable using SSB. Recently I received a report on an interesting QSO, not only using QRP SSB but also completely solar powered. Dave Starkie, G4AKC, of Blackpool, reports:

I am very active on the HF bands, mainly pedestrian mobile, but some bicycle as well, however I am delighted to report that on 22 April at 11:30Z, I had a 2-way QSO with John, ZL2JBR, in Wellington New Zealand on 20m SSB whilst using only solar power from my pedestrian mobile two- wheel trolley (without the use of any batteries) with a 5/3 report. The QSO was witnessed by Lee, G0DBE.

The radio was a Mizuho MX-14S 20m 2-watt SSB/CW handheld transceiver powered by a double solar panel on the trolley. The antenna was a full-size quarter wave, again mounted on the trolley, the ground was a homemade tuned frame ground tuning unit without radials.

"Fortunately, it remained sunny throughout

the QSO. If the sun had disappeared then so would have my transmit and receive! I operated from Lytham on the Lancashire coast; very close to the sea water."

Pedestrian beach portable

Peter Parker, VK3YE, is an avid QRP operator who actively promotes home-made equipment and has his own style of operating pedestrian mobile. Peter spent his youth at rubbish tips, taking apart donated radios and TVs and building electronic projects that mostly did not work. Early projects were built with screws and breadboards inspired by George Dobbs' Simple Electronics book. Attempts at soldering failed until he realised that it really needs new solder rather than that reused from terminal strips in valve radios.

Milestones included the construction of a crystal set in 1980, discovering shortwave on a valve receiver in 1981 and building a simple breadboard 'electronic organ' in 1982. Hours were spent putting wires into springs on a Tandy 150-in-1 electronics set. Amazingly some wires could be pulled out and the project would still sort of work with only half the parts in circuit. Two back to back AM/shortwave radios led to the discovery of amateur SSB activity and a novice licence. A one valve CW transmitter provided the first contacts on 3.579 MHz. The value of frequency agility was an early lesson. Many VFOs were built but most drifted. Equipment on more bands and modes were later constructed. A move to a beachside suburb near Melbourne, Australia, brought increased interest in HF portable and pedestrian mobile operating. The *Beach 40* is a well-known 40 metres DSB (Double Sideband) transceiver of his own design. Peter maintains a popular YouTube channel and writes books on a good range of amateur radio topics. His website is at http://www.vk3ye.com

Nonchalantly carrying his gear in bags over his shoulder, when pedestrian mobile he uses an extended fishing rod to allow him to have a decent and relatively compact antenna for whatever band he has in mind, typically operating on 40 metres. His operating from the local beach allows him to have a decent path away from the coast, especially on CW.

Peter Parker VK3YE

VK3YE pedestrian mobile on HF

QRP portable - in a bottle!

Pete Juliano, N6QW, has been continuously licensed for over 61 years, thus he can legitimately be called an OM. An Extra Class License, since 1977, he had an opportunity to be DX while stationed on Midway Island as KM6DD during 1963/64. A degreed Electrical Engineer, his retirement provided an opportunity to enjoy hardware level homebrewing. Many of his projects have been published in various ham publications and span all things QRP from an SSB Transceiver in a Champagne bottle, to a Shirt Pocket SSB Transceiver and is now working on homebrew Raspberry Pi QRP SDR Transceivers. One of his most ambitious undertakings was a solid state QRP Version of the Collins KWM-2 which he calls the KWM-4. Many of these projects are currently shown on his QRZ.com webpage. In 2015 N6QW was inducted into the QRPARCI Hall Of Fame. He also likely holds the world record "for the numerous times he soldered his fingers together!".

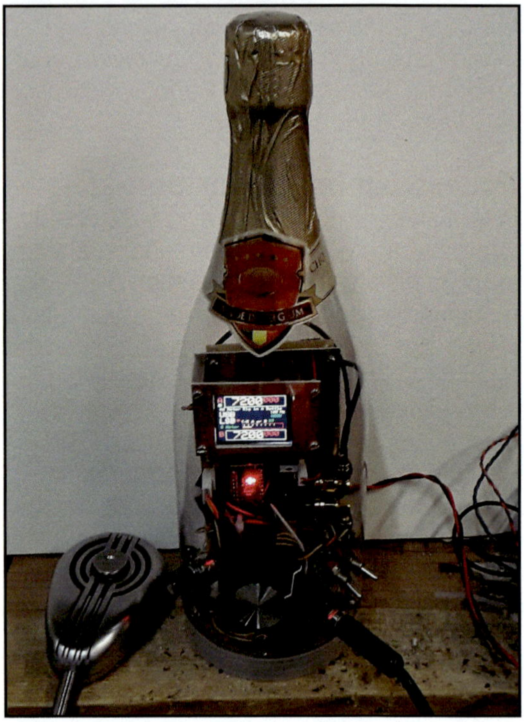

QRP in a bottle!

To share with the greater ham community his many projects N6QW maintains several ham radio related websites, takes part in the famous SolderSmoke Podcast with Bill Meara N2CQR and a separate site dedicated to his other homebrew passion – Cooking. N6QW can be contacted at n6qwham@gmail.com

Pete Juliano N6QW

Chapter 9

Basic Operating Tips

Once a station has been assembled, the would-be QRP operator has to venture onto the bands. Initially this is quite a daunting thought. The amateur bands can be tough, most stations will be running twenty times more power and may be armed with a better antenna. Can this little station give any account of itself on the bands? Judging from what I hear daily, the answer is – a surprising amount. All that is required is a little effort, a little skill and few ground rules. So how to proceed?

The first thing is to define success. As K8ZT once said: "Radio is a hobby not a job". We are in it for fun. Some will want to simply enjoy casual operating, contacting whom they can and being pleased with it. Others may have a more competitive edge and want to set standards and goals. One way is to compete against oneself and keep a countries-worked tally, or a distance-against-power record. The GQRP Club offers a series of awards for the QRP operator (see end of book). From the USA are the QRP ARCI '1,000 miles per Watt ' and the ARRL 'DX-CC QRP' awards.

Use your operating time wisely

The wise QRP operator attempts to understand and follow some of the basic rules of radio propagation; which frequencies work best and when. There are good beginner's books on propagation, so find one and read it, trying to grasp the basic terms and concepts. There are a few basic good rules.

A good start is to use the highest band that is open. This is sometimes called 'MUF working' (maximum usable frequency). Forecasts of the MUF can be found in propagation tables and some web- sites. Operating at, or close to, the MUF reduces path loss. A basic rule of thumb is to use the higher frequency bands (14MHz and higher) during the daylight times and the lower frequency bands (10MHz and below) during the dark hours. Even this basic rule has variants. An example is that on the 14MHz path to the USA, the "your sunset / his sunrise" period seems to work well. But because some ionospheric layers seem to follow the path of the sun very closely, it is often easier to work the first weak USA signals in the early evening, than the much louder signals later in the evening.

Other 'rules of thumb' include the winter being best for east-west contacts and the summer best for north-south contacts. Old hands, who use directional antennas, often point their beams to the east in the morning, the south at the middle of daylight, and the west in the evening. If the QRP operator has a fixed antenna with a better 'take-off' in a particular direction, this is useful information.

Sometimes signals from a particular direction may appear to be very strong, indicating a good path to that direction. Some long-distance paths show optimum propagation conditions in only one direction. An experienced operator will not only listen for which signals are strong but make a note of the stations these signals are contacting. The best outgoing path may not be the one with the strongest incoming signals. It is well worth mastering the basics of propagation to avoid wasting time on what may be fruitless frequencies.

A very useful beginners guide to the subject of radio propagation is Radio Propagation Explained by Steve Nichol, G0KYA. It gives an easy to digest introduction to the vagrancies of radio wave propagation at frequencies from short wave to UHF.

Help is also at hand from the Web. The radio propagation page from the Radio Society of Great Britain is available at https://rsgb.org/main/technical/propagation/ This page provides a comprehensive listing of Internet sources relating to radio propagation at all levels from introductory to professional.

Going on the air

It is often said that the best DX operators, listen 95% of the time and transmit 5% of the time. Learning how to listen is an essential tool for the QRP operator. The secret is to go on a band and listen for a while before trying to make any contacts. Ask questions like: Which stations are strong and where are they located? Who are these stations contacting and where are they located? How strong might my signal be with louder stations? If your station receiver or transceiver has an attenuator switch, try switching it on to get an indication of how weak an incoming signal becomes. If you can- not hear the signal with the attenuator switched in, that station will probably not hear your signal.

Avoid switching on, finding a clear spot and calling "CQ". In fact, experienced QRP operators hardly ever call "CQ". They know that more contacts are likely to be gained by 'tail-ending'. Tail- ending is quickly calling one of two stations already in contact when that QSO draws to the end. It requires a little skill and practice to get right but it is a useful way of gaining contacts when using low power. Tail-ending needs to be quick and slick, on CW usually just sending "<station callsign> DE<own callsign> KN" immediately after the last transmission has occurred.

If you know who began the QSO, choose that station from the two stations in contact; courtesy says that it is his/her frequency. It is tempting just to call the louder station of the two. Tail-ending is much easier when both stations can be heard clearly and when they conclude the contact with the correct "VA"; "end of work" procedure signal. But many stations add an extra "73" greeting or a "dit-dit" at the end of a QSO. Experience and individual judgement are both required.

Do not maintain marginal contacts for too long. If the signal report from the other stations indicates that your signal is being read with difficulty, restrict the contact to a basic exchange and a courteous thank-you. The other operator is probably doing you a favour in trying to maintain contact in difficult circumstances. His signal report, your location and your name are probably enough. The other station may not welcome struggling to find out what your weather is like and what equipment is in your station.

When using CW (Morse code) accuracy is better than speed. Stick with the speed at which you know fewer mistakes are likely to be made. The secret is to make your Morse easy to copy. If the other station is sending rather faster than your optimum speed, do not be tempted to increase your sending speed. Remember that one of the standard ways of repeating missed information is to send it more slowly.

Use simple exchanges of information. The first exchange is the important one; it gives the basic information to the other station. A typical first exchange may be:

XXXX DE G3RJV TNX CALL UR RST 579 579 QTH ROCHDALE ROCHDALE NAME GEORGE GEORGE HW? AR XXXX DE G3RJV KN.

The other station now knows his report, your location and your name, which is quite enough for the

first exchange. Notice I did not include the power, or anything about QRP. That can add an "S" point to his signal report! If stations know you are using low power they do tend to give lower signal reports.

Depending upon his report of your signals, decide how much information to give in the next exchange. You may be able to maintain a long QSO or it might be appropriate to cut your losses and move on.

Calling CQ

Although many QRP operators eschew calling "CQ", sometimes that is the only option and some- times it can be fruitful. I never go seeking other stations on a band before I have tried calling "CQ" on the QRP Calling

Frequency on that band.

The International QRP Calling Frequencies (**Table 9.1**) are the places to meet other QRP opera- tors. They are always glad to contact other low power stations and these frequencies have been agreed amongst the major QRP organisations as the places to seek two-way QRP contacts.

The best way to call "CQ" on these frequencies, and indeed on any frequency, is the simple 'three times three' pattern: CQ CQ CQ DE G3RJV G3RJV G3RJV K

Listen for a few seconds, perhaps tuning slightly either side of the frequency. Use receiver incremental tuning (RIT) if it is available. It is then a good idea to send "QRL?" ("Is this frequency in use?") and begin the 3 x 3 call again. Long rambling CQ calls add nothing to the procedure and will probably annoy other operators. Do not send the CQ calls on CW at speeds above your competent operating speed.

When calling CQ on one of the QRP Calling Frequencies it can be useful to indicate that your station is a QRP station. After the final "G3RJV" of the 3 x 3 call, I might add "QRP". Some overseas stations will be heard sending their callsign with the suffix /QRP (e.g. XY7ABC/QRP). "QRP" is not a legal suffix for use by a UK amateur radio station. Do not send "G3RJV/QRP" but rather "G3RJV QRP".

Use correct procedure

When the receiving station is listening to a weaker station, the task is made easier by accurate Morse code sending and the use of correct procedure signals. It is simple politeness to use the accepted Morse abbreviations. The main ones that help the QSO to proceed smoothly are shown in **Table 9.2**.

So, the basic rules are:

- Send accurate rather than fast Morse code.
- If using telephony (for example SSB) always use the accepted international phonetics when spelling out a word.
- Use correct procedural abbreviations.
- Send only the necessary basic information in the first transmission: station report, name and location (QTH).
- Send extra information (power, equipment, antenna, weather etc.) on the second transmission.
- Do not prolong the QSO if conditions are 'marginal'.
- Always thank the other station for the contact.

TABLE 9.1: INTERNATIONAL QRP CALLING FREQUENCIES (kHz)

	160m	80m	40m	30m	20m	17m	15m	12m	10m
CW	1836/1843	3560	7030	10116	14060	18086/18096	21060	24906	28060
SSB		3690	7090	—	14285	18130	21285	24950	28365

TABLE 9.2: COMMON MORSE CODE PROCEDURE SIGNALS IN AMATEUR RADIO

K	Invitation to transmit (general)	Often used at the end of a CQ call
KN	Invitation to transmit (to a specific station)	Between exchanges
AR	End of message	During a QSO
SK or VA	End of transmission. Same – both sent as one string	To finish a QSO
Also, commonly used, and useful, are:		
HW?	"How do you receive me"	During QSO
R	"Received all sent"	During QSO

Contests

There are a number of specific QRP Contests held annually, which offer good opportunities to enjoy low power operating with a competitive edge. Some of these are listed later in the book. Several major international contests also offer a QRP section, with alternative awards and diplomas for operators using low power. Most of these are published in the RSGB's RadCom magazine.

QRP operators have also learned how to use major contests for working new and unusual countries. They use contests to seek new countries, rather than entering them as a competitor. It is sur- prising how well the big contest stations can hear a low power signal if it will gain them a few extra points.

A fruitful way to do this is to join in a contest as it draws to the end. Most of the big contest stations have worked most other stations in the event and are keen to contact any new stations to gain more points. But remember, in major contests you will be dealing with 'no-nonsense' operators. They will expect the standard information exchange that is appropriate to that contest and no more. Find out the appropriate contest exchange from the contest rules or listening to a few contest exchanges. It will probably involve the use of a serial number. Do send all the required information or the contestant's efforts will be wasted.

For contest, and general, operating it is very useful to have a listing of country prefixes to identify the location of stations heard or contacted. The ARRL DXCC List is very popular, especially with

those QRP operators who are chasing the ARRL QRP-DXCC Award. The RSGB publishes the comprehensive RSGB Prefix Guide for the identification of stations from their prefixes.

Operate a low powered station with care and confidence, and viable and enjoyable contacts will be made. Operate as if you expect to gain QSOs and enjoy compliments for the results of your low powered signal.

Finally – What is 72?

Listening around the QRP Frequencies will yield stations which send the greeting "72" at the end of a contact. 72 is the QRP version of the "73" greeting. It was first used by Oleg Borodin, RV3GM at the end of the 1980s and has caught on as a greeting between QRP stations.

For that matter, what is "73"? The folklore varies a little, but it is said that the old telegraphy operators used to send two dashes, six dots and then two dashes as a greeting before sending their messages. Split in the middle this forms the numbers 7 and 3. "73" has remained in the amateur radio community as "best wishes".

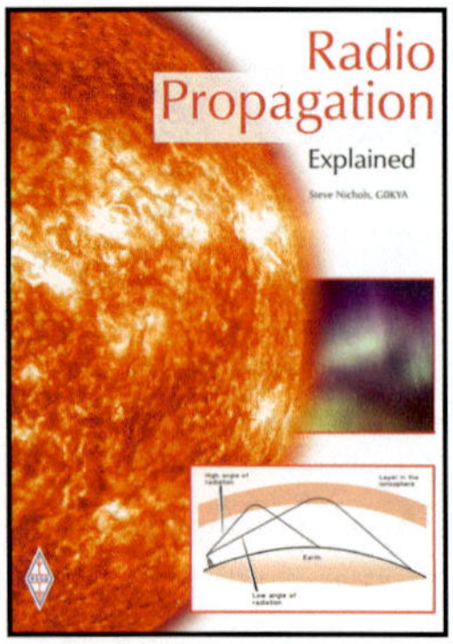

Useful operating books for the QRPer

Chapter 10

QRP Antenna Ideas

There is often an assumption that QRP operators must be using large, or expensive, antennas to achieve any worthwhile results. In fact, the opposite is more often the case. Within the philosophy of many QRP operators is the assumption that they are going to make their own antennas. Most QRP operators use homemade wire antennas, many of them configured to make the best use of limited space. This means assessing what can be done within the confines of the home situation and learning from the experience of others. To meet one very experienced "antenna-man" I can commend the excellent (and still available in Jan. 2021) antenna website of L B Cebik, W4RNL(SK), at www.antentop.org/w4rnl.001/radio.html.

This book is no medium to do justice to antenna choices for the operator. It would take another complete book and I would not be the author. My best advice is that the reader buys, or borrows, some of the many good books written on simple practical antennas for amateur radio use. One of my favourites is the GQRP Club Antenna Handbook. It is full of practical antenna ideas, all of which have been used by QRP operators and published in the SPRAT journal.

Another favourite of mine is W1FB (SK)'s Antenna Notebook, a book which deals simply with antenna theory and practice and suggests good working examples for the QRP operator. It contains a wealth of information written in the lucid W1FB style. The book is now out of print, but may available second-hand from various web sources, and is available as a PDF download from: https://www.pdfdrive.com/w1fb-qrp-notebookpdf-e21048255.html.

The RSGB offers a good variety of antenna related books, many of which are suitable for the beginner and the practical antenna builder. Three examples are shown here.

Backyard Antennas, a book by Peter Dodd, G3LDO (SK), explains how, by using a variety of simple techniques, it is possible to achieve high performance from a compact antenna in

limited space. The book covers end-fed and centre-fed antennas, rotary beams, loops, tuning units, VHF/UHF antennas, antenna and mast construction, transmission lines, and how to estimate and measure performance.

Another book by Peter Dodd is Building Successful HF Antennas, in which he explains what makes an effective HF antenna, how to build one and how to measure its performance.

A more recent publication from the RSGB is Successful Wire Antennas, edited by Ian Poole, G3YWX and Steve Telenius-Lowe, 9M6DXX/PJ4DX/KH0UN. It covers a vast array of wire antenna designs including dipoles, doublets, verticals, loops and end-fed wires, as well as subjects such as impedance matching, ATUs, baluns, feeders, antenna masts and rigging.

The half-wave dipole

Perhaps the simplest common wire antenna is the half-wave dipole. Roughly speaking it is a wire of half a wavelength at the required frequency, split in the middle, and fed with 50-ohm coaxial cable. It requires no antenna tuner as an accurately cut

half-wave dipole will match the 50-ohm feeder, which in turn will match the usual 50-ohm output impedance of a transmitter or transceiver.

That is 'roughly speaking' for in reality a dipole should be cut somewhat shorter than an electrical half-wave because of end effects. There are several factors that affect the resonant frequency of any antenna. These factors include the height above ground, the diameter of the wire, nearby structures, the effects of other antennas in the area and even the conductivity of the soil. Fortunately, there is a standard formula that can be taken as a starting point in cutting wire for a dipole. For a centre fed, wire

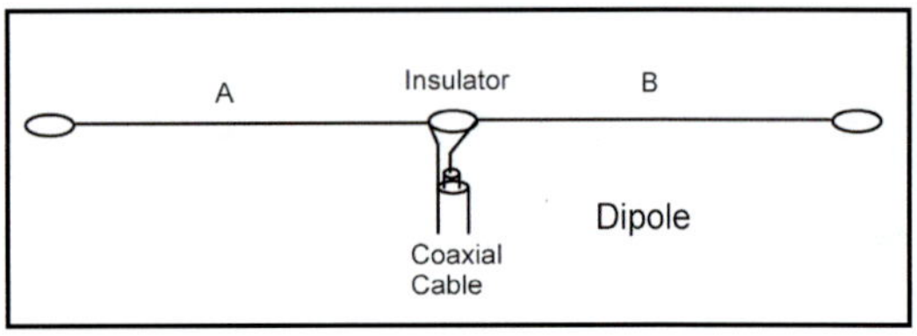

Fig 10.1: The Half-wave dipole

dipole, this formula is length in feet = 468 / frequency in MHz. It is a good idea to cut the wire a little too long and tune the dipole to the required frequency using an SWR Bridge or antenna analyser.

Making a dipole is simple as **Fig 10.1** shows. The two legs of the dipole (A and B) are made from wire a quarter-wavelength long, or slightly shorter as we have seen above. A good insulator is necessary in the centre to hold the two wires apart so that connections can be made to the inner wire and outer screen of a length of coaxial cable. The best results will come from good quality 50-ohm coax cable like RG-58/U or RG-8/U. The coax feeder can be of any reasonable length.

Specially made dipole 'T' centres can be bought but I have tended to use homemade centre insulators made from fibreglass or some other good insulating material. For quickly built dipoles, I have used double 'chocolate block' (terminal block) connectors. If the dipole is to hang outdoors, the centre connections do need waterproofing. It is also usual to add insulators at either end of the wire legs.

As with all antennas, the secret is to mount the dipole as high as possible and as clear as possible from nearby large objects. For the length formula above, the dipole ought to be mounted in a horizontal plane. **Table 10.1** shows worked examples for a practical half-wave dipole at some QRP frequencies.

TABLE 10.1: HALF-WAVE DIPOLE LENGTHS

Frequency	Total Length	Length each leg
7.030MHz	66.57ft / 20.29m	33.28ft / 10.14m
14.060MHz	33.29ft / 10.15m	16.64ft ./ 5.07m
21.060MHz	22.22ft / 6.77m	11.11ft / 3.38m
28.060MHz	16.68ft / 5.08m	8.34ft / 2.54m

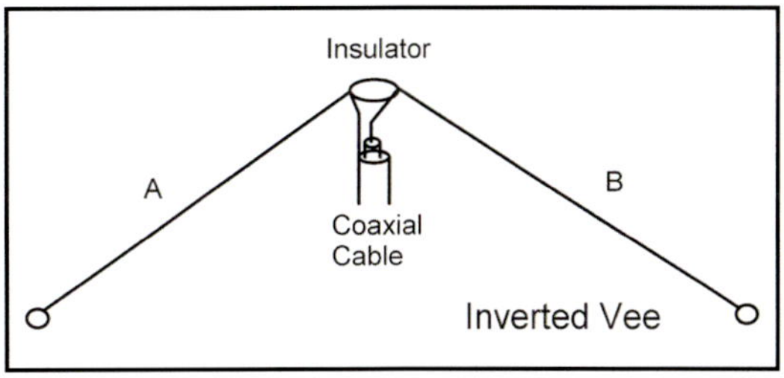

Fig 10.2: The Inverted-vee dipole

TABLE 10.2: INVERTED-VEE DIPOLE LENGTHS

Frequency	Total Length	Length each leg
7.030MHz	63.91ft / 19.48m	31.96ft / 9.74m
14.060MHz	31.95ft / 9.74m	15.97ft / 4.87m
21.060MHz	21.33ft / 6.50m	10.66ft / 3.35m
28.060MHz	16.01ft / 4.88m	8.00ft / 2.44m

The inverted vee dipole

The inverted vee is simply a dipole in which the legs are mounted at an angle (**Fig 10.2**). This configuration can be very useful if only one high fixing point is available from which an antenna can be hung.

For many years I lived in a three storey Edwardian house with a high gable end and hung many antennas from this accessible high point.

By sloping the sides of a dipole, it is also, possible to fit it into a smaller space. Placing the legs at an angle of about 90 degrees does add a surprising amount of available "leg length" to a dipole.

The legs of an inverted vee dipole are cut about 3 - 5% shorter than a dipole at the same frequency. 4% is commonly used as a constant in the calculation of inverted vee dipole lengths. **Table 10.2** lists some lengths for the QRP frequencies.

A multiband dipole

It is also possible to make a multi-band dipole by feeding more than one dipole antenna from the same coax cable (**Fig10.3**). This is often referred to as a "Fan Dipole".

The requirement is to separate the individual legs of the dipole either side of the common feeder points. This can be done by means of a system of insulated spacers.

The idea is open to individual experimentation and antenna fans have built many inventive examples. Some practical examples of multi-band dipoles use ribbon cables of parallel conductors separated by plastic sleeving.

The electrical characteristics of the dipole legs will be changed by the proximity of the adjacent legs. This does mean individual pruning and tuning of the lengths of each leg. With a little care, such arrangements can make a good compromise antenna that will give acceptable results over several amateur bands.

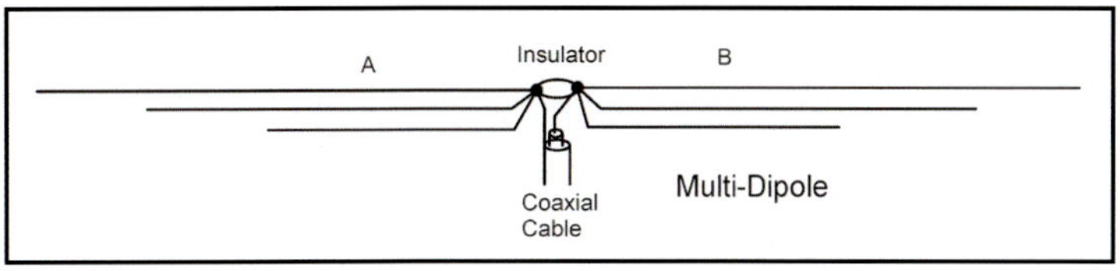

Fig 10.3: A multi-band dipole

The doublet antenna

A dipole antenna can be fed with higher impedance open-wire feeder; sometimes called twin or balanced feeder (**Fig 10.4**). An open-wire feeder comprises two parallel wires, held apart by insulating material. This might also be described as "ladder line".

Many amateur antenna constructors make their own open-wire feeder by mounting spacers between two wires. This type of feeder has an impedance in the order of 300-400 ohms. I am a great fan of open-wire feeder. It offers far less radio frequency loss to the signal and is more versatile when used with a balanced

antenna tuner (see later in this chapter). I commonly used a variant called Slotted Ribbon Cable available in the UK from WH Westlake Electronics (see Traders Appendix).

A dipole fed with open-wire feeder is usually called a Doublet antenna and has the advantage that, by using a balanced antenna tuner, it can work on several bands. My usual antenna at the home location is an open-wire fed 105ft (32m) doublet, which

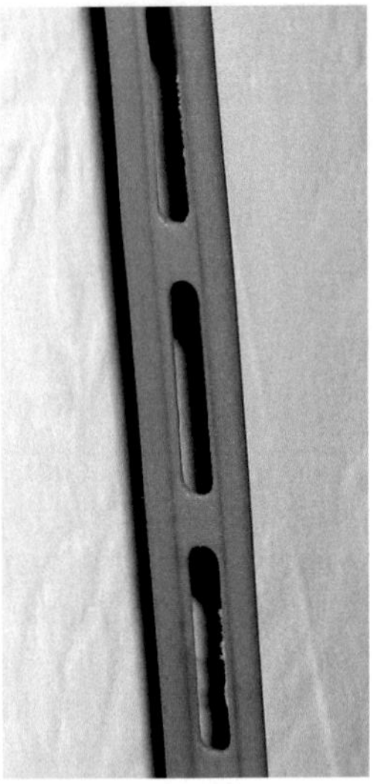

Slotted Ribbon Cable

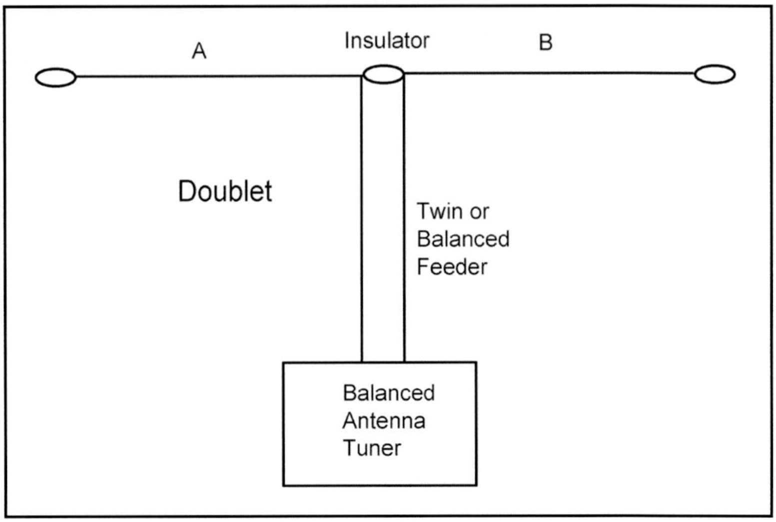

Fig 10.4: Doublet antenna

I tune for all amateur bands with a Z-Match antenna tuner.

Readers wishing to try the doublet approach should refer to the many books with articles on the subject. Open-wire fed antennas are very popular with QRP operators. When I first began amateur radio, some antenna advice I received said: "put up a wire, as high and as long as you can, split it in the centre and feed it with open-wire feeder".

The end-fed wire antenna

When I was a boy in the 1950s, almost every house seemed to have a 'radio pole' at the bottom of the garden. This was in the days of valve radios that needed an aerial or antenna. The aerial often took the form of a length of wire running from a downstairs room to the eaves of the house and then down the garden to the pole. These were 'end-fed' antennas and that was the first type of antenna I used for my amateur radio.

Fig 10.5 shows a typical end-fed antenna. A piece of wire goes from the station to a remote high fixing point. In this case the wire gains its initial height by being pulled up towards the roof using some insulated line. Strong nylon line or kite line can be used. The fixing point for the far end could be the classic 'radio pole' or a tree or some other tall building or fixture.

The end-fed wire has two main drawbacks: it requires an antenna tuner to match it to the transmitter and it does require a good ground (or earth) system. A wire that is a multiple of a half wavelength of the frequency in use has a

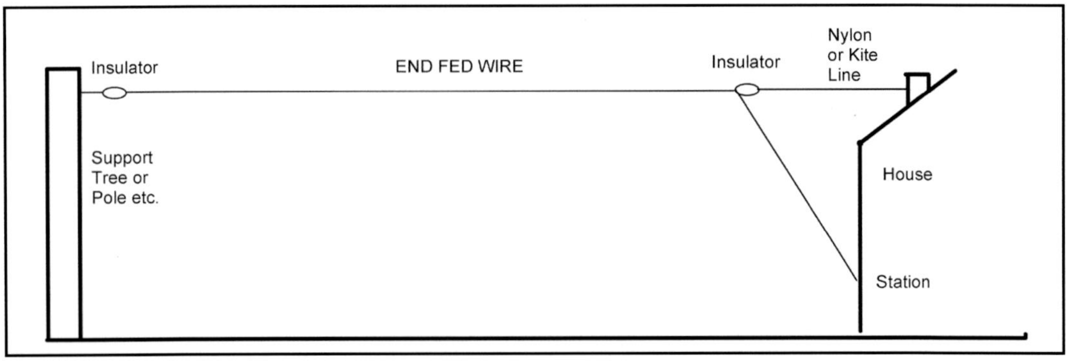

Fig10.5: End-fed antenna

high impedance at the feed-point. Wires under a half wavelength have a low impedance. What is required is a tuner that can handle a great range of impedances likely to be found at the transmitter end of the wire if it is being used over a range of bands.

The L-match

A simple form of antenna tuner that is often used for end-fed wires is the 'L-match'. This is so named because it only has two elements, an inductor and a capacitor, and these are arranged in an 'L' con- figuration. For practical purposes of tuning, both the inductor and the capacitor should be variable. **Fig 10.6** shows an L-match antenna tuner configured to match a high im-pedance wire and a low impedance wire. The version shown in (A) will match a longer (high impedance) wire and the version in (B) will match a shorter (low impedance) wire.

One way of practically doing this would be to simply turn the tuner around according to what kind of matching is required. In the past I have built version of the L-match in which the capacitor (C1) could be switched to either end of the inductor (L1).

C1 and L1 are variable. The classical compo-nent for L1 would be a roller-coaster inductor; a variable inductor in which a slider moves along the turns of the coil to vary the inductance, but these can be expensive, and even rare, com-ponents. Acceptable results can be had from a conventional

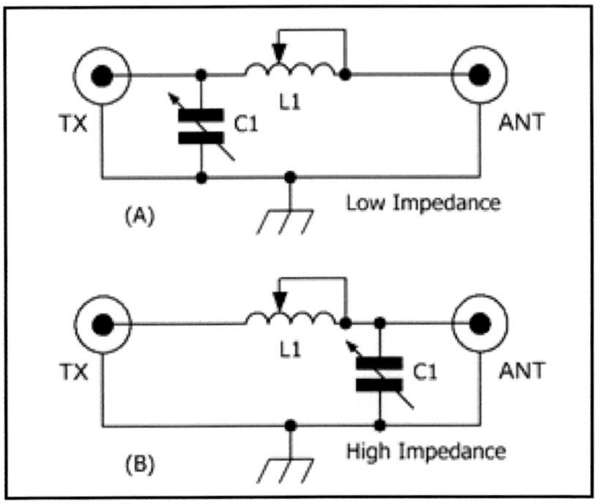

Fig 10.6: L-match antenna tuner

inductor with a switch to select the number of turns included in the circuit. For a typical HF bands L-match this could be 60 turns of 22SWG (21AWG) enamelled copper wire wound on a 1in (25mm) diameter cylindrical former. A 12-way wafer-type switch could be used to select the number of turns included in the circuit. The coil of the inductor will need to be wound with 'tapping points' to allow the selection of turns. The variable capacitor should be in the 300pF to 500pF range. The photograph shows an L-match antenna tuner that I built some 20 years ago. This one uses a roller-coaster inductor.

A switchable L-match tuner

This is a commercial inductor in which the windings, in bare copper wire, fit into grooves on the former. The principle is the same when using enamelled copper wire wound side by side on a former.

A 12-way wafer switch (right) selects taps on the coil. Notice that the tapping points are stag-gered. They begin, on the left with a switch step of 10 turns and get smaller until the last tapped point is only 2 turns from the previous tapping.

This is because smaller changes in induct-ance are required at the higher frequency bands. The central switch allows the variable capacitor to be placed at either end of the inductor to allow either configuration in Fig10.6.

A complete, practical L match tuner, using a toroid instead of an air-cored former, is shown in **Fig 10.7**. The inductor, L1, has a 12-way, 1 pole, wafer switch to select a range of inductanc-es. The 340pF variable capacitor is a polyvaricon type. Values up to 500pF would be suitable. Polyvaricon capacitors (sometimes called sol-id-dielectric) are avail- able from several suppliers or may be found in older AM portable transistor radios. If the L-match is to be used at powers above 10W an air-spaced variable capacitor will be required. Extra capacitance can be added to C1 by switching in C2 with a switch (S3).

The variable capacitor (C1) can be switched to the transmitter or antenna side of L1 with a centre-off toggle switch, S2, S1 is wired so that in the first position it selects no turns of the inductor connecting the transmitter directly to

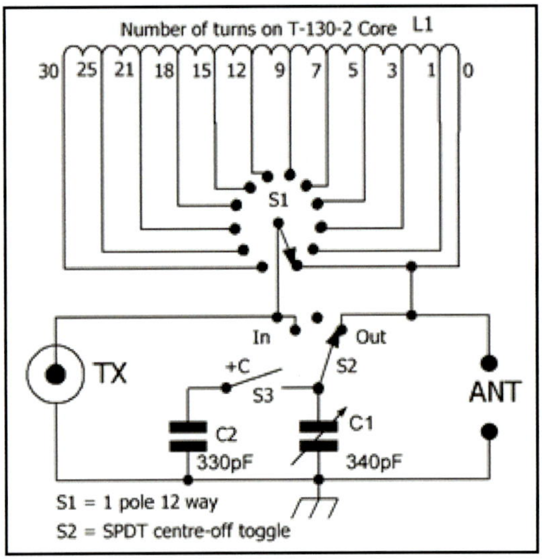

Fig 10.7: A practical L-match antenna tuner

A roller-coaster inductor

A switchable L-match tuner

the antenna. The problem is that C1 is normally connected to L1. The centre-off toggle switch (S2) in the centre position removes C1 from the circuit. The signal can then pass directly through the circuit, eliminating the tuner.

The tapped coil is wound on T-130-2 toroidal core. The coil has a total of 30 turns giving a total inductance of around 10 microhenries; this is sufficient to tune from the 3.5MHz band upwards. The coil is wound with 24SWG enamelled wire with taps at 0, 1, 3, 5, 7, 9, 12, 15, 18, 21, 25 and 30 turns. The tapping points are made by pulling out a loop wire, about 3cm from the core, and making two or three twists to secure it in place. The tapping wires are scraped with a knife or glass paper until the copper is exposed and tinned with solder using a hot soldering iron. L1 is mounted directly to the back of the 12-way switch soldering the tapping wires to each of the twelve switch terminals. Identify the switch tag for the first switch position then work around the remaining switch tags, soldering each tap to the next tag.

When first using the tuner, approximate settings can be found using received signals. Set C1 about half way and rotate S1 for the loudest signals. Peak the signals by rotating C1 and this will give the approximate tuning point. Final tuning can then be achieved with a low-level transmitted signal and a standing-wave bridge.

Ground or counterpoise

End-fed antennas are tuned against ground (earth). It is notoriously difficult to obtain a reliable ground connection. Many radio amateurs have laboured to drive a series of long copper rods into their gardens, some have buried sheets of chicken wire and I know of one who buried a scrap car in his vegetable garden . . . all in an attempt to get a lot of metal to contact with a lot of ground.

A better alternative is to use a 'counterpoise wire'. This is a length of insulated wire running close to, or laying on, the ground. Ideally the counterpoise wire should be a quarter-wavelength for the band in use. I have had very acceptable results using an end-fed wire for several bands with a single counterpoise wire that is a quarter wavelength for the lowest band in use. The calculations for the counterpoise can be taken from Table 10.1 earlier in this chapter.

Chapter 11

Commercial QRP Options

Perhaps the easiest way to become a QRP operator is to buy a commercial QRP transceiver. Off-the-shelf QRP transceivers have been available since the end of the 1970s and several transceivers are in current production. Second-hand QRP equipment can also be found for sale in the small advertisements in the back of amateur radio magazines, at amateur radio events, in the journals of QRP organisations and in web-based selling formats such as eBay and some QRP websites. Since new products are being made available all the time, I can only include items currently on sale at the time of writing. Other very popular QRP items will be included in the section on kits.

Yaesu FT-818 Transceiver

Based on the very popular Yaesu FT-817 and FT-817ND, the FT-818ND, it is probably the best-known QRP transceiver currently available. The basic specifications are quite impressive:

Size: 5.3 x 1.5 x 6.5 inches (includes space for 8 'AA' cells or Yaesu NiCD battery)

Weight 900g (~2 lbs)

Transmitter Coverage: 160-10m HF Bands, 5 MHz band frequencies, 6m, 2m, 70cm VHF/UHF

Power Output: 6W SSB/CW/FM, 2.5W AM (on external 13.8V supply or 2.5W on

The Yaesu FT-818 transceiver

internal batteries)

Operating Modes: USB, LSB, CW, AM, FM, W-FM, Digital (AFSK), Packet (1200/9600 FM)

Digital Modes: RTTY, PSK31-U, PSK31-L, and user defined USB/LSB (SSTV, Pactor, etc.)

Receiver Coverage: 0.1-56MHz, 76-154MHz, 430-470MHz

The FT817/ND and now the FT818ND have been responsible for introducing many radio amateurs to low power operation. Many are attracted to buying the FT-818 because it packs so many things into such a small space, and has an improved specification to the FT817. A complete

amateur radio station covering all the bands from 1.8MHz to 430MHz, using all the common modes and including a general coverage receiver "that could be posted through a letter box", to quote G3TUX. I know of sever- al radio amateurs who bought the FT-817 as a 'second station', or a holiday radio, and found it gave them so much fun, that it has become their main station.

The FT818 makes it possible to own a very compact and versatile QRP station. They come complete with a hand microphone for telephony modes, and also includes an internal Morse keyer, which requires only a paddle to be added for CW operation. For serious CW operation the optional 500Hz CW Filter is probably essential as the basic receiver selectivity is quoted as 3kHz @ 6dB down, which is rather wide to resolve CW signals on a crowded band.

This transceiver has built up quite a reputation as a QRP portable transceiver. The full specification would fill two more pages of this book so for more information readers can refer to the Yaesu website (http://www.yaesu.com) or the main dealers.

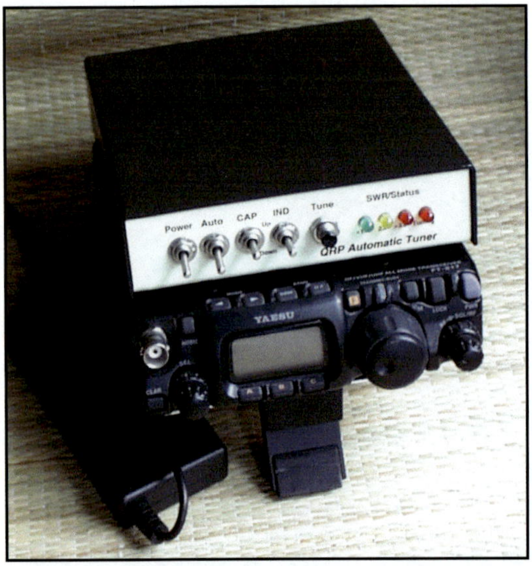

The picture shows my compact FT-817 station. Beneath the transceiver is a small switched-mode 12-volt power supply capable of running the FT-817 to full output. On top of the transceiver is an LDG QRP automatic antenna tuner, which will match the transceiver into a variety of antennas. The LDG Z11 Automatic Tuner has now superseded this antenna tuner

MFJ QRP transceivers

MFJ Enterprises is a company perhaps best known for a wide variety of amateur radio accessories and antennas but they have produced QRP equipment and kits for some years. Many of these are still available and work well. Recently MFJ have supplemented their own units by importing and selling the Chinese Xiegu G90 portable low power (20w) transceiver

MFJ is an interesting company. Founded in 1972 and still run by Martin F Jue, MFJ began operations in a small rented hotel room in Starkville, Mississippi. The first product was an active audio filter; an out-board unit to aid the CW selectivity of receivers. The company then moved to a rented room in a large building and thence to a large trailer. MFJ eventually occupied a former roller-skating rink. Most of the company is now housed in a large unit on an industrial estate.

I have a fondness for MFJ. I owned one of their original audio filters and used it with my first Argonaut 505 transceiver. In the 1990s I took my (then) schoolboy son to a radio convention in

Dallas, Texas. Our GQRP Club stand was located across from the MFJ stand. My rather 'bored by radio' son joined Martin Jue's equally bored daughter for an afternoon at a local amusement park.

MFJ 90XX series QRP CW transceivers

The 90XX series is a range of no-frills single band transceivers. The last two number of each model indicate the band: MFJ-9040 - 40 meters; MFJ-9030 - 30m; MFJ-9020 - 20m; MFJ-9017 - 17m and MFJ-9015 for 15m. Features include:

- Superhet receiver
- 750Hz crystal filter
- Fast recovery AGC
- RIT control with centre-detent
- 5-watt RF output (variable)
- Built-in speaker
- Low current consumption: receive - 50mA, transmit - 1A
- Size 162 x 60 x 150mm
- Weight 840gm

The MFJ-90XX series make ideal introductory transceivers to low power operation or can serve as holiday or portable stations. Tim Allison, G0TYM, owns both the MFJ-9040 and MFJ-9020 trans- ceilers and together with the matching MFJ-971 Antenna Tuner and MFJ Portable Power Pack, they give him a 'take-anywhere' MFJ combo. Tim finds the ATU effective in matching temporary wire antenna systems and the 9040 and 9020 can be quickly swapped over by changing power and ATU leads.

He told me of a holiday trip to the Isle of Mull. He set up a W3EDP wire antenna between a bedroom window and a tree at the end of the garden with a counterpoise wire running along the bedroom floor. Arriving on a Saturday evening, he was able to quickly set up the station and have a QSO with DL9AAS to test the station before taking his evening walk along the beach. Radio operating during the week on Mull was limited to short periods in the mornings and evenings, before and after enjoying exploring Mull. The week of occasional operating on 40 and 20m yielded twenty countries throughout

MFJ-9015 CW Transceiver

Europe and some enjoyable longer QSOs with UK stations. Tim did not rely on the superior output power or antennas of other stations to secure his contacts. He used the

International QRP frequencies to find other low powered stations including YU1CA and QRP stations in France, Austria, Germany, Spain and the UK.

Further information on the MFJ90XX series can be had from the MFJ website at http://www.mfjenterprises.com. The transceivers are stocked in the UK by Waters and Stanton plc.

MFJ have recently added to their CW QRP "cub" transceivers with the MFJ-93xx numbering range that once again have the single band replacing the xx in the number and they are available as a kit "K" suffix or fully wired "W" suffix so there are models MFJ 9340K (40m CW QRP kit transceiver) MFJ 9340W (40m CW QRP transceiver, fully built). In this range there are also the 9320K, 9320W, 9317K and 9317W.

See the next chapter for more details on the MFJ Cub kits.

The G0TYM MFJ portable combo

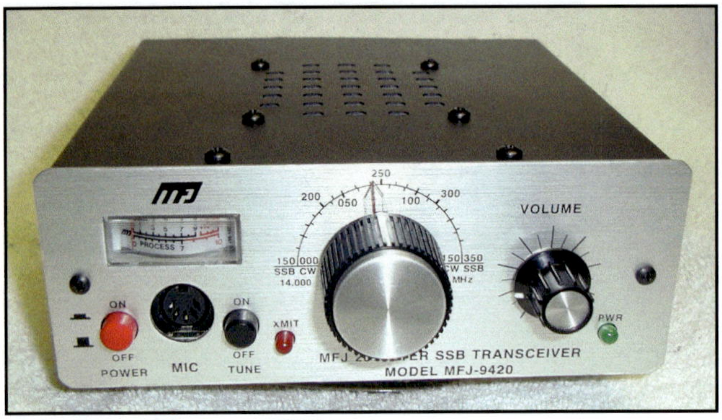

MFJ-9420 (20 metres SSB)

The MFJ Cub

MFJ-94XX series QRP SSB transceivers

MFJ Enterprises manufacture a range of single band QRP SSB transceivers. The series is: MFJ-9402 2-meter SSB Transceiver, MFJ-9406 6m SSB Transceiver, MFJ-9410 10m SSB Adventure Radio, MFJ-9420 20m SSB Travel Radio, MFJ-9440 40m SSB Travel Radio, MFJ-9475 75m SSB Travel radio. There're also kits, the MFJ 9340, for example, for the 40 m band

Specifications vary a little from model to model, but the main features are:

Single band - single conversion superhet transceiver

IF selectivity -6dB at 2.5kHz

AGC - audio derived

Audio: more than 1 watt into 8-ohm speaker

Tx current 2 amps peak (13.8 volts)

Receive current 50 - 100mA

Output 10 watts PEP

RF compression processor, 600-ohm microphone

Analogue meter reading S-units on receiver and power out / ALC on transmit

CW available as an add-on module

Size 162 x 60 x 150mm

Weight 850gm

My only experience of the MFJ-94XX series is being at the receiving end. Some years ago, I was a passenger in the Kanga Products van driving through Europe towards an amateur radio event in Germany. We were running a mobile SSB station, albeit with 100 watts to a 'Texas Bugcatcher' antenna and enjoying a variety of contacts. The only contact we had with a low power station was with a GQRP Club lady member in Wales using an MFJ-9420 Transceiver.

I regret to say I have forgotten her callsign and name but I do clearly recall being in contact with a pleasant easy-to-read SSB signal. Nor do I recall the signal report I gave but the MFJ-9420, using a wire antenna, it was clear enough to copy in a noisy van on a busy road in Luxemburg. Again, an excellent little transceiver for portable or casual use or as a first QRP SSB station.

XIEGU transceivers

These Chinese made transceivers are available through dealers around the world and represent the new influx of lower priced Chinese made HF transceivers in the 20w and lower range of output. As mentioned above the XIEGU G90 is marketed through MFJ in the US.

The G90 replaces the earlier X108G model and the G1M, the earlier X1M model.

XIEGU G90

- A short-wave transceiver with amateur radio bands: 0.5 ~ 30 MHz, operating modes: SSB/CW/AM/FM, transmission power up to 20W.
- [Detachable Display Unit] - The display

unit and host can be separated so that you can flexibly arrange the location of the host.

- [Built-in Automatic Antenna tuner] - With a high-power automatic antenna tuner, you can quickly set up the antenna.
- [Clear reading screen] - All information is clear at a glance in the display. Special waterfall indicator are used to view the bands.

XIEGU X5105

Xiegu X5105 is an ultra-portable HF and 50 MHz QRP transceiver with a general coverage receiver, large LCD display, automatic antenna tuner and a built-in high capacity battery making it the perfect portable companion. All mode 4.5 watts output 160m – 6 metres.

(This is XIEGU's challenger to the Elecraft KX-3 although at a lower power level – see later)

XIEGU G1M

This is the XIEGU's entry level rig with 5 watts SSB/CW output. It is an SDR with general receive 0.5-30 MHz and transmit capabilities on 80, 40, 20 & 15m radio bands. It's highly portable and has an OLED screen with S-meter and spectrum display.

XIEGU also make an CE-19 interface unit to allow the use of these rigs with any amplifier and indeed produce their own amplifier the XPA-125B but this is definitely outside the QRP realm.

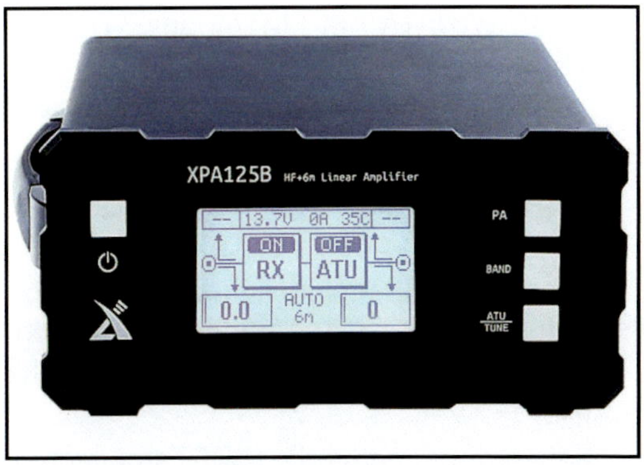

XIEGU XPA-125B amplifier

XIEGU G90

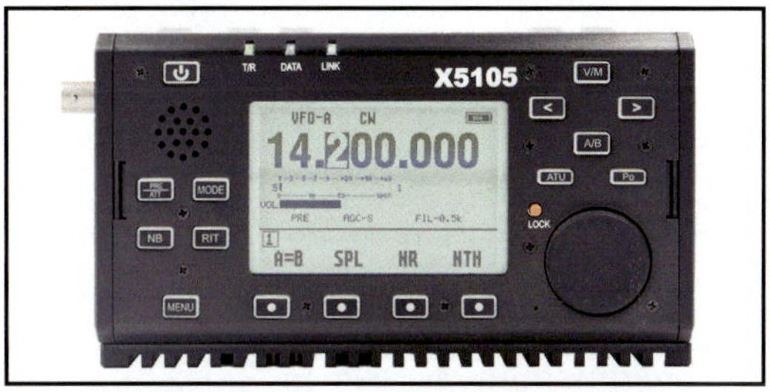

XIEGU X5105

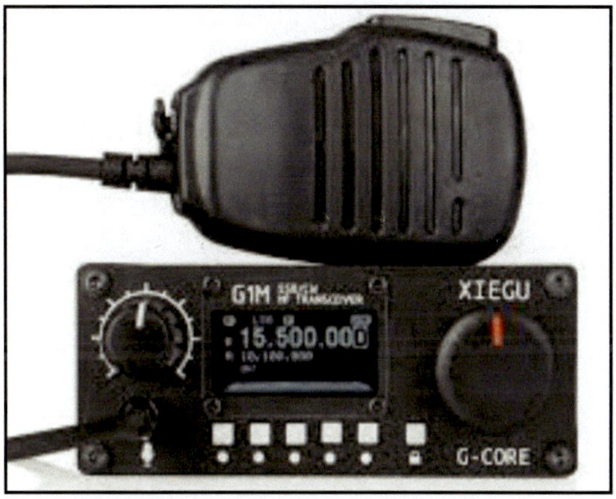

XIEGU G1M

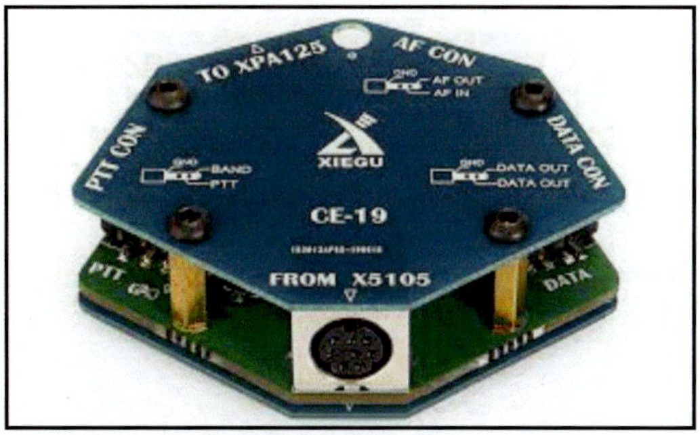

XIEGU CE-19 interface

YouKits

The HB-1B MK5 QRP transceiver

The HB-1A ultra-compact 7-band CW QRP transceiver was designed and manufactured in the city of Nanjing, China by BD4RG. It was a 'trail friendly' radio offering the 40, 30, and 20 metre bands. The HB-1A was advertised to put out about 5 watts with 12-volt DC power supply and 2 watts nominal from 8 internal AA batteries.

The Chinese manufacturers are now called You Kits, who manufacture the latest version, the HB-1B, also features continuously variable selectivity in conjunction with a crystal filter, and offers the 80, 60, 40, 30, 20, 17, and 15 metre bands. Other interesting features include 30 memories, RIT, automatic CQ with user's call sign, attenuator, tuning steps of 10Hz, 100Hz and 5kHz, IF filter bandwidth from 400Hz to 3kHz, built-in electronic keyer, and a clear LCD display. An external 12-volt supply gives up to 5W output and 14 volts (maximum) gives up to 7 Watts.

HB1A and HB1B QRP transceivers

The receive drain current is only about 60mA. There is also provision for using three LS14500 lithium cells. The size (132 x 85 x 35 mm) and weight (380gm) make the HB-1B ideal for portable or back- packing use.

The HB-1B is available from the number of traders in the UK, including Waters and Stanton and Nevada. Other QRP radios are planned by You Kits. QRPers should keep their eye on the webpage at http://www.youkits.com/.

HB-1B 4 Band Trasceiver

Other Chises Manufacturers

There are other Chinese makes and models, not covered here but worth a look are:

Lab 599 SDR in a pre-cast solid aluminium case for rough portable use.

KN-990 SDR all mode HF 15-watt output with full spectrum scope

McHF/RS-918/HH-918 SDR 160-10m 10 watts with all modes including FreeDV digital Voice

Elecraft

The American company Elecraft has gained a fine reputation for their range of transceivers. Most of them are sold in kit form and will be considered later in the kit section of this book. They currently have two QRP offerings the second, the KX2 only comes in ready built form but the KX3 is available as either a kit or ready built. The first to come to market was the KX3 and at the time was the most sophisticated (and expensive) QRP transceiver on the market. Elecraft describe it as an ultra-portable, all-band/all-mode high-performance transceiver for home, mobile, or field use. It took the portable operations market by storm.

The Elecraft KX3 & KX2 QRP transceivers

The specifications of the KX3 are very impressive if the following brief summary is anything to go by:

160-6 metres (2m with KX3-2M module)

SSB/CW/DATA/AM/FM modes; 10 W+ PEP (100 W with KXPA100 amp)

1.7" x 3.5" x 7.4" (4.3 x 8.9 x 18.8 cm); only 1.5 pounds (0.7 kg)

Current drain as low as 150 mA; internal battery (8 AA cells, any type)

Advanced DSP provides dual watch, noise reduction, noise blanking, auto notch, stereo audio effects, audio peaking filter (APF)

Software-defined radio (SDR) architecture plus optional roofing filters

The Elecraft KX2 QRP transceiver

Operates in data modes (PSK31/RTTY) with or without a PC

Includes RX I/Q sound-card outputs; works with PC RS232 or USB ports; free firmware upgrades. Options include and internal wide-range antenna tuner, internal NiMH battery charger, dual roofing filter module, attached high-performance keyer paddle, internal 2-metre module, compact hand mic with UP/DN buttons, 100-W amplifier and ATU.

More details can be found from the Elecraft web site at http://www.elecraft.com/KX3/kx3.htm

The Elecraft KX2 QRP transceiver

The KX-2 was an even smaller unit with a built-in microphone so that the unit could be held and used like a military "walkie-talkie"

In order to reduce size and complexity there is no support for 160m or 6m and no 2-metre module. This is truly a QRP all mode HF hand-portable unit.

The specifications for this unit on the Elecraft website state:

All modes: SSB, CW, Data, AM and FM weight 13 oz. (0.3 kg)

80-10-meter ham bands; general coverage receive from 3-32.0MHz

Internal 2.6 Ah, 11 V Li-ion battery option

Internal wide-range ATU option

Built-in microphone and speaker

DSP providing dual watch, noise reduction, noise blanking, auto-notch, stereo audio

SDR architecture

Operates in data modes (PSK31/RTTY) with or without a PC

Size: 2.8"H x 5.8"W x 1.5"D;

High-contrast LCD

Power output up to 12 W on 80-15 m, 8 W on 12 and 10 m (typ.)

Use with built-in mic or MH3

CW keyer with 8-50 WPM range

Built-in digital voice recorder (DVR)

Three CW/DATA message memories

PIN-diode T-R switching

DSP speech processing

ICOM IC-705.

ICOM impacted the entry level HF transceiver market with their IC-7300 160-6m (4m in the UK & Europe) 100w SDR transceiver in January 2016. It had a large touch screen and a spectrum scope along with other advancements and took the market by storm. In June 2020 they had managed to shrink most of this unit (without the ATU) down into a 5/10w portable version in direct competition to the Elecraft KX-3 ultra-portable transceiver. It is an understatement to say it was another success. Although heavier than its opponents, the fact that it added 2 Metres, 70cm and D-Star to the specifications of the IC.7300 (although it can't do 4 metres), made it a hard choice to beat if you were in this $1500-ish market for a QRP portable transceiver. Many amateurs complained that it came without an internal ATU but this was soon addressed by after-market external units and indeed later by an external unit from ICOM themselves as well. Many QRP portable operators use resonant antennas and so an ATU is not needed but it seems many people still would like to have one and, on this point, the Elecraft KX3 and KX2 have an advantage with their built-in options. The unit can run with an internal battery but then with reduced output (5w), for the full 10w the IC-705 is capable of, an external 13.8v supply is required.

Launched at the same time as the transceiver was a small ruck sack "LC-192" with a pocket designed to take the IC-705 but many chose not to but this due to the price asked for it and the availability of other OEM products to protect and carry the 1.1 KG rig.

Here are the full details from the Icom IC-705 marketing brochure:

- **IC-705 HF/VHF/UHF Portable Transceiver:** The new portable HF/VHF/UHF IC-705 has many great features such as SDR platform, internal battery, GPS, Bluetooth and D-STAR, all in a compact and lightweight body. The IC-705 uses the same 4.3 inch colour touch screen display as the best-selling IC-7300 featuring real-time spectrum scope and waterfall display.
- **HF/50/144/430 MHz Multimode Operation:** From HF to 50/144/430 MHz, you can enjoy a variety of bands in D-STAR DV, SSB, CW, AM and FM modes. The IC-705 receives continuously from

The ICOM IC-705

30 kHz through the 144 MHz band. You can also enjoy FM broadcast and Airband reception.

- **RF Direct Sampling System:** The IC-705 uses the RF direct sampling method. This method greatly reduces distortion. Down-conversion IF sampling method is used for 25 MHz and above
- **Real-Time Spectrum Scope and Waterfall Display:** The IC-705 inherits the performance and functions of the highly-acclaimed IC-7300 scopes. You can visually see band condition and check for clear frequencies.
- **Large Touch Screen Colour Display:** The large 4.3-inch colour TFT touch LCD is the same size as the IC-7300 and offers intuitive operation of the IC-705's functions, settings, and various operational visual aids, such as the band scope, waterfall, and audio scope functions.
- **Compact and Lightweight Design:** "Base station radio" performance and functions are packaged in a compact size of approximately 20 cm, 7.9 in (W) × 8.35 cm, 3.3 in (H) × 8.2 cm, 3.2 in (D). The weight is approximately 1.1 kg (including its BP-272 battery). It can be held in one hand.
- **BP-272 Li-ion Battery Pack or 13.8 V DC External Power Supply:** The IC-705 utilises the high capacity Li-Ion battery from the ID-51A and ID-31A handheld radios. A 13.8 V DC external power supply can be used for operation and for charging the BP-272.
- **Maximum Output Power 5 W (BP-272), 10 W (13.8 V DC):** In portable mode, the IC-705 has a maximum output power of 5 W from the BP-272 which can last approx. 3 hours. This is perfect for true 5 W QRP as well as the 0.5 W QRPp operations. Once you connect an external 13.8 V DC power source, you have up to 10 W.
- **Supports Connections Using WLAN/Bluetooth® Technologies:** Utilises WLAN/Bluetooth® technologies for linking and remote control. The VS-3 headset (optional) enables more comfortable operation via Bluetooth®
- **GPS Functions:** An internal GPS receiver and antenna enhance your operations by providing location logging, RX/TX locations via D-PRS®, "Near Me" repeater search/scan, QSO recording with metadata, and internal clock synchronization.
- **microSD Card Slot:** The microSD card slot enables the storage of user-profiles, QSO recording, TX voice memory keyer, RTTY logging, GPS data, screen capture, firmware upgrades, and programming.
- **Micro USB Connector:** In addition to audio and rig control, the micro USB port allows charging, making digital mode operation easy with a minimal number of cables.
- **D-STAR Functions:** The latest G3 Gateway and DV mode features are supported on the IC-705. You may set-up direct access to the D-STAR network with Terminal/Access point modes. Additionally, the IC-705 has the Photo Share feature so that you can share photos without the need of a computer.
- **Programmable Speaker/Microphone:** The supplied HM-243, programmable speaker/microphone has user assignable function buttons to allow frequency and volume adjustments at the tip of your fingers, without having to access the transceiver front panel.

Chapter 12

Kits for QRP Operation (UK)

QRP stations are often a rich mixture of all kinds of equipment. Some use only commercial equipment, others are completely home built and many contain a mixture of both. My mother used to say, *"You can't beat cooking from fresh ingredients"*. Many QRP home constructors do the radio equivalent by choosing, or designing, a circuit, gathering the necessary parts and assembling the project. Others opt to build their equipment by buying a kit - a sort of 'QRP ready meal'.

QRP appears to be the branch of amateur radio best endowed with kits for the radio constructor. There is a lot of choice for the would-be kit builder. Many of them are supplied by small companies, in some cases just a sole trader producing kits as a spare time occupation. The variety is amazing; from kits for the absolute beginner to kits which enable the construction of equipment which is technically superior to many commercial products.

I have sometimes heard experienced amateur radio constructors being disparaging about kits, suggesting that it is little more than 'painting by numbers. My view is that kits can be an excellent way into the world of radio construction. One of the most off-putting aspects of radio construction is having the first constructional project fail to work. Sadly, many beginners have not progressed beyond their first endeavour.

By buying a kit, the would-be constructor is embarking upon a project, which should, if the kit manufacturer is worth anything, result in success. Another frequent problem for beginners, and indeed seasoned constructors, is getting hold of the required components. With a kit, it should come complete with all the necessary parts. The do-it-yourself constructor has to decide how to build the project. The only information may be a circuit diagram and a list of parts. Some will choose to design and etch their own printed circuit board or they may use one of several alternatives to a printed circuit board. Most kits are supplied complete with an already etched circuit board which may even have a silk-screen overlay showing the positions for all the components.

As with most endeavours, the best plan for the beginner is to choose a simple kit; a simple QRP transmitter or a useful station accessory. Several kit manufacturers produce modest kits, which can be completed in an evening, or a weekend, which do not cost much and should work first time. For a small outlay in money and time it is possible to mount the first step towards becoming a builder of QRP equipment. Later in this book, I describe some basic principles of workshop set-up and practice that apply to kit building as well as do-it-yourself construction. This chapter gives a brief tour of kit manufacturers and their products in the UK. In the subsequent chapter I will describe what is available in the USA.

Generally, there has been a difference between what kit manufacturers supply in the UK and in the US. British kit manufacturers have tended to supply, a printed circuit board, the electronic components required to build a working project and leave the constructor to find a case or housing for the completed project. In the US, most kit producers have supplied everything including a suitable case, very often with a silk screen front panel. Both approaches are valid. It is useful and convenient to have a ready-made housing for the equipment, which can make the finished item look good. Others contend that finding a box and drilling a few holes in it should not be a problem. Very often hardware is expensive and a case and knobs can sometimes cost more than all the electronic content. The do-it-yourself approach to hardware can make the kit much cheaper. There is also a certain amount of satisfaction in adding the personal touch to a kit. It makes the project individual, whereas a supplied housing makes the completed project look the same as everyone else's completed kit. In recent times, however, UK suppliers have been offering cases and several US suppliers have produced basic kits without cases.

Walford Electronics

Tim Walford, G3PCJ, runs Walford Electronics from his farm in Somerset. Tim produces an interesting and unique set of kits with their own design and construction philosophy. The kits range from simple projects suitable for a beginner to the Bristol, a modular multi-band, multi-mode transceiver. As is usual the kits come complete with all the electronic components and a printed circuit board. They differ in that the kits are supplied with a front panel for 'open construction'. The front panel is another piece of printed circuit board material with the legends for the controls etched on the front. Tim Walford says, "I shall be disappointed if you can't find something in my range that you would like to build! There is a wide range of designs, with something for most construction abilities. These kits are for people interested in analogue radio - so no micro-processors or software defined radio here! You can start with a small project or build your own complete station - most projects will suit the practical examination aspects of obtaining your licence. The circuitry is what I concentrate on with the aim of maximising 'performance' for whatever you decide to spend."

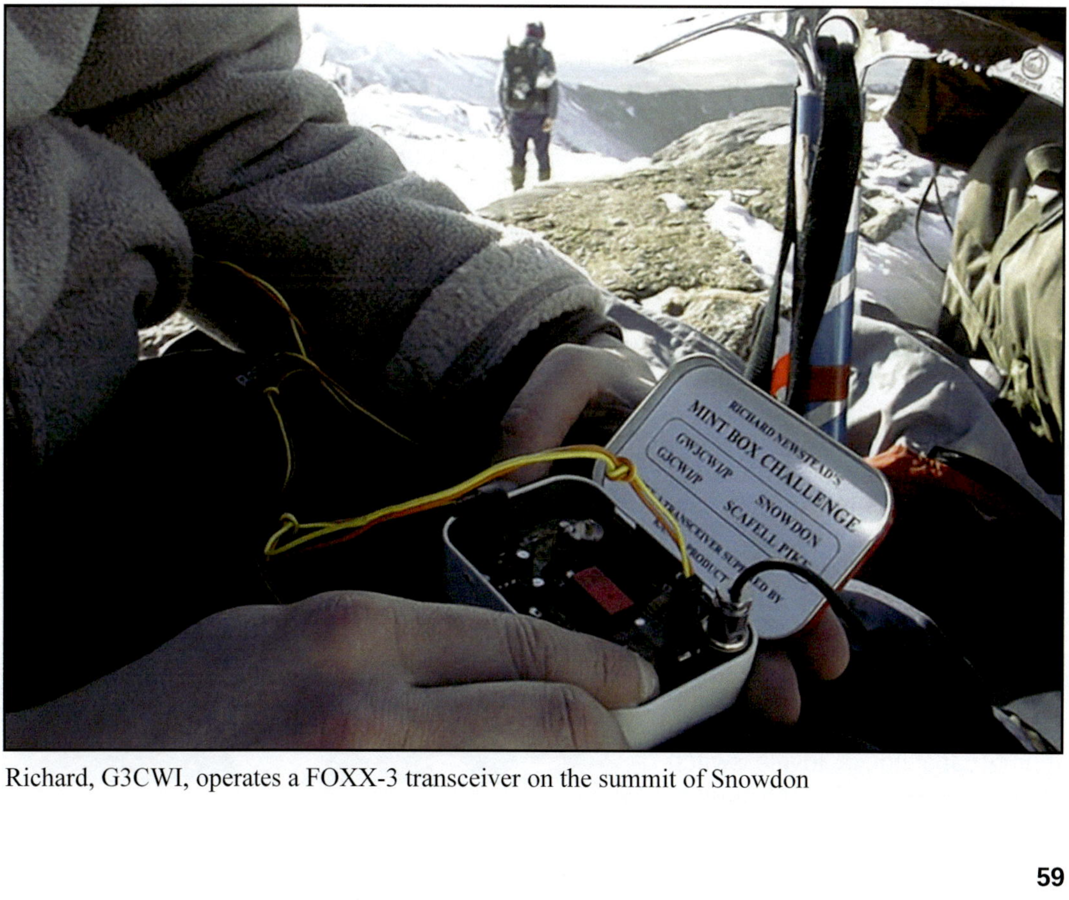

Richard, G3CWI, operates a FOXX-3 transceiver on the summit of Snowdon

An example of a Walford Kit: the Walford Culm transceiver

The range is too large to list here and changes frequently with new products. Details of the current kits can be had from the Walford web site: www.walfordelectronics.co.uk/. Here is just one example. The Culm CW transceiver is a builder friendly design with several advantages over earlier versions of direct conversion projects for those wanting to work CW. It is normally supplied for 80m using a ceramic resonator VFO for coverage of the CW section of the band, but also it can also be used on other bands up to 20m using a crystal for that particular band instead of a ceramic resonator. It has a direct conversion receiver with narrow 725 Hz bandpass audio filtering as well as a humped 725Hz low pass filter. The output can drive 8-ohm loudspeakers or 'Walkman' type 32-ohm headphones. It has a small PCB front panel with full sized potentiometers for the audio gain (volume) control and fine-tuning controls, the main polyvaricon tuning capacitor, a toggle switch for the receiver incremental tuning, and sockets for the loudspeaker and key.

The fine-tuning control doubles as the receiver incremental tuning which is automatically removed during transmission, or by the RIT switch. Keying is semi-break in, with relay change-over of the aerial circuits. The rig includes receiver muting during transmission and a sidetone oscillator, with a separate pre-set for its level in the loudspeaker. There are double tuned receiver band pass filters and double half wave transmitter output low pass filters. Transmitter output is 1.5W on any of the bands chosen, when using a 13.8V supply.

The MFJ Cub transceiver

The MFJ QRP-Cub Transceivers are a series of single band CW transceivers which use surface mount technology to achieve a pocket-sized package. The kit version has all SMT parts ready mounted and soldered. The kit builder inserts and solders the through-hole parts. The Cub was designed by K1BQT, a member of the QRP-ARCI hall-of-fame. There are versions for 80, 40, 30, 20, 17, or 15 metres. The output power is variable from zero to 2 watts on models up to 20m and 1W on 17 and 15m.

The MFJ Cub

The receiver covers a 50kHz segment of the band, which is adjustable to suit the operator. Other features include sharp crystal filtering, 0.2 microvolt receiver sensitivity, audio derived AGC and full break-in changeover from transmit to receive. The small size (90 x 47 x 98mm) and low current drain (36mA on receive, 380mA on transmit) make it ideal for portable use. The Cub is also available fully assembled. Details can be found at https://mfjenterprises.com/search?q=MFJ+Cub

QRP Labs

QRP Labs was founded by Hans Summers G0UPL, a very active experimenter, builder and designer of QRP systems. Digital modes have expanded from CW to include both further data modes and digital speech. The company is very active in QRP. One of their main products for QRP CW is the QCX+ kit, described as a feature-packed, high performance, single band 5W CW transceiver kit with WSPR beacon and built-in alignment and test equipment. Available for 80, 60, 40, 30, 20 or 17m bands.

The kit is stated as being easy to build, with a plated through hole PCB, giving approximately 3-5W CW output (depending on supply voltage), with a recommended supply voltage of between 7 to 16 volts, with a class E power amplifier which runs cool with no heatsinks. It has a 7-element low pass filter to ensure that there is compliance with regulatory bodies and no spurious transmission outside the band of use. It employs CW envelope shaping to remove key clicks, and has a high-performance receiver with at least 50 dB of unwanted sideband suppression. It has a 200 Hz CW filter with no ringing, and employs a Si5352A synthesised VFO with rotary encoder tuning. It permits the use of an Iambic- or straight key through an option in the firmware. Other key features include pre-set frequencies and offset CW adjustment with VFO split A/B operation. It has also an on-screen S meter. A useful feature is that it has built-in test equipment in the form of a signal generator along with alignment tools which include a voltmeter, RF power meter and frequency counter. Finally, it also has a CAT control interface, all-in-all a very comprehensive kit.

Hans is currently developing a new QRP transceiver, the multi-mode and multi-band QSX. It's sure to be an interesting kit when it arrives.

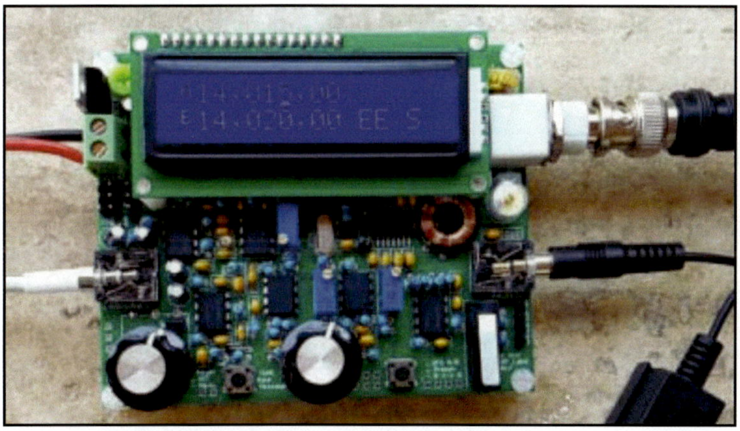

QRP Labs QCX+ kit

Chapter 13

Kits for QRP from the USA

The USA is a rich source of kits for the QRP enthusiast. It is probably easier to be a manufacturer of kits in the USA because of the larger internal market and some American QRP kit producers make their sole living from their kit sales. In recent years some of the larger American QRP clubs and organisations have produced their own kits. The kits produced by QRP groups tend to come and go, so I will only mention the main commercial producers of kits in the USA.

Obtaining kits from the USA

Most of the American QRP kit manufacturers will supply kits outside of the USA, many of them accept payment via credit cards. This involves greater postage costs and the goods will attract customs and VAT charges, usually at the time of delivery. Almost all, American kit producers have their own websites, which usually include information about overseas ordering and payment.

If you live in the UK, some American kits can be obtained within the UK where such charges will already be included in the price. I have mentioned the American kits available through Waters and Stanton Ltd. (https://hamradiostore.co.uk/) for example. There is a wider range of American kits (including Elecraft) available through QRP project, (https://www.qrp-shop.biz/) a German company, in Berlin, run by Peter Zenker, DL2FI (SK) & Nikolai DL7NIK where additional customs charges will not apply to those living within the EU however of course postal and customs duty may again apply with any kit exported from the EU, e.g. to the UK. Check all charges before purchase.

Elecraft

I recall being in the in the QRP hospitality suite at the Dayton Hamvention when Elecraft first showed off the K2 Transceiver. It was a new concept for the QRP equipment constructor; a fully featured HF transceiver, in kit form, that looked and performed like the top end of the commercial market. It certainly impressed the QRP fans. Many orders were taken on that first weekend on the basis of seeing what was not, at that stage, a fully working model.

The Elecraft K2 transceiver

The basic K2 is a QRP CW transceiver with features which include; Twin VFOs with selectable Band Up/Band Down; Direct frequency entry and ten memories; Independent receiver/transmitter tuning (RIT/XIT); variable VFO tuning rate; 'S' meter; power meter; Digital frequency readout; Variable output power (0.5-10 watts); CW operation on eight bands (10/12/15/17/20/30/40/80 metres); Full SSB receive mode and 12V operation (10-15 VDC). The Basic K2 offers QRP CW operation with up to 15 watts output. It can be built as a K2/100 by including the internal KPA100 option to give 100 watts output. Other options

The Elecraft K2 transceiver

include SSB, the 160m band, a high-performance noise blanker, computer control and audio filter options

Building such a complex receiver may seem a daunting task but Elecraft commend it to the beginner. The constructional manual describes all the stages and no specialist test equipment is required beyond a digital multimeter. In fact, the kit is configured to include its own test equipment: a built-in general-purpose frequency counter, digital voltmeter and ammeter. Elecraft will even point kit buyers to K2 builders who will build the kit for a fee.

The K2 has certainly built up a huge reputation and there is a range of K2 logo merchandising and several websites featuring delighted K2 owners. It has become rather a cult QRP rig. The K2 kit is still available in 2021 from Elecraft 24 years after its first introduction.

Pacific Antenna

This company is situated in Fayetteville in Arkansas, and a good range of kits can be seen on their website at https://www.qrpkits.com/ covering classic designs from radio amateurs, including the Survivor, which is a 75-meter SSB/CW kit designed by famed QRP designer Steve Weber, KD1JV, and the BitX series of QRP designs by Ashhar Farhan, VU2ESE. As an example of one of these, the Survivor produces around 10 W PEP at 13.8v on the 75m band, and has an impressive receiver sensitivity of 0.2 mV. It has a 5-pole crystal ladder filter, with a tuning range of either 325 or 175 kHz which is selectable, with coarse and fine-tuning controls. It operates in SSB, CW, and "tune" modes. The kit includes a silk screen plated through hole PCB, aluminium chassis, and needs only 2 amps at 13.8 volts from a suitable power supply.

HF Signals (Ashhar Farhan VU2ESE)

The BitX20, which also comes in a 17m version, is designed as an SSB transceiver, with a typical power output of around 10 watts.

There is also another version available from HF Signals, the company founded by Ashhar Farhan. VU2ESE to employ out of work women who were on the bread-line. Called the mBitX, this is multiband transceiver covering all the HF amateur bands. It produces 10 watts except for on the 10-metre band, where it falls to around 5 watts. The mBitX employs an Arduino Nano controller. The original concept of the BitX series of transceivers was that the design need not use components which were difficult to obtain and/or expensive. The design also includes innovative bidirectional circuitry, which allows transmission and reception signals to be sent along the same internal circuit pathway.

HF signals can be found on the web at https://www.hfsignals.com/

The Survivor transceiver

The BitX20 transceiver

Hendricks QRP kits

Doug Hendricks, KI6DS, has over 15 years' experience designing kits, having started with the NorCal 40 in 1993, and has produced over 20,000 kits through his association with the NorCal QRP Club and Hendricks QRP Kits. The company offers a wide range of QRP kits that can be viewed at the pacific antennas site - http://www.qrpkits.com/ Here is one example kit; the PFR-3 transceiver.

The Hendricks PFR-3

The PFR-3 (Portable Field Radio) is a three-band self-contained transceiver kit covering 40, 30 and 20 meters. It is self-contained in that the transceiver, antenna tuner, memory keyer and eight AA batteries are contained in one small case. External power can be supplied through a separate jack, and the user can switch between batteries and external power from a front panel switch. There is one add-on option available; the PFR paddle that plugs directly into the front of the transceiver.

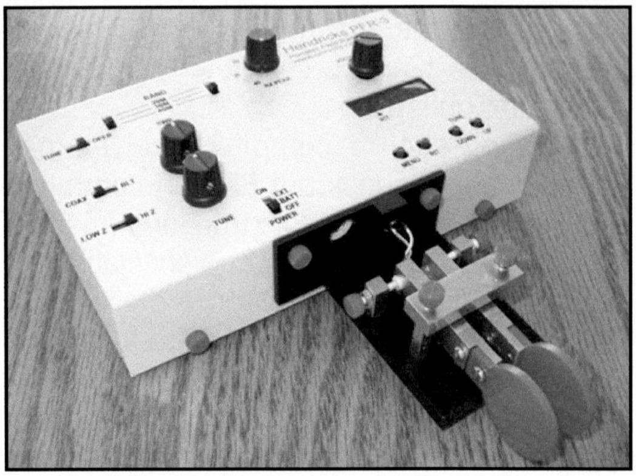

The Hendricks PFR-3 transceiver

The PFR-3 is designed by Steven Weber, KD1JV, and the custom enclosure is from Ken Locatable, WA4MNT. It includes digital read out (four digits), a DDS VFO and microprocessor control. Firmware updates for the PFR is provided free for life. Parts for this kit are through-hole, with some pre-mounted SMD components. The PFR-3 has become popular with those who enjoy working QRP portable.

Chapter 14

Simple Stations & Accessories

How does an intending QRP operator get on the air with low power? The obvious answer is to buy a new or used QRP transceiver, or a transceiver kit. But there are simpler, and cheaper, ways to sample low power operating. May I suggest some of the basics of setting up a QRP station?

A simple transmitter with a receiver

When I began my amateur radio, most people operated with 'separates'; a separate transmitter and receiver. The usual approach was to buy, or build, a receiver and then add a homebuilt transmitter. I guess that I was a radio amateur for about ten years before I owned a transceiver; a combined transmitter and receiver.

An easy way into QRP operating is to use an existing receiver, or the receiver portion of an existing transceiver, and build a basic QRP transmitter. Many QRP operators have begun with a simple crystal-controlled transmitter and a receiver. They are easy enough for a beginner to build and the satisfaction of contacting stations with a small project you have built yourself is a real amateur radio experience.

A little transmitter board I had a lot of fun with was the Pebble, designed by Doug DeMaw. He was kind enough to give me his prototype on one of my visits to his workshop. After Doug's death from leukaemia in 1997, I brought out the board, dusted it, and used it for about a month in memory of an old friend. The three watts from the Pebble was combined with my vintage Davco DR-30 receiver to give me a lot of worthwhile CW contacts on the 40m band.

Operators used to more sophisticated modern equipment may question the viability of a crystal- controlled transmitter, where the frequency of transmission is fixed by the single frequency of a crystal. What if there are no stations to contact at that frequency? It is possible to contrive the crystal oscillator as a VXO circuit (variable crystal oscillator) where capacitance, and sometimes inductance, is used to pull the crystal a little from its nominal frequency. However, the frequency range of a VXO is quite limited without causing stability problems.

By carefully choosing the crystal frequency, it is possible to 'park' the transmitter on a fruitful frequency for contacts. A crystal on a QRP

The Pebble Transmitter by Doug DeMaw, W1FB

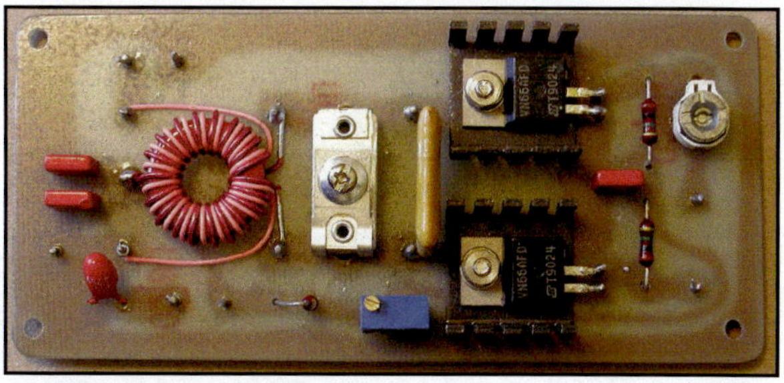
The NAT (Not Another Transmitter) by George Burt, GM3OXX. An example of a very simple 5W transmitter for the 80m band

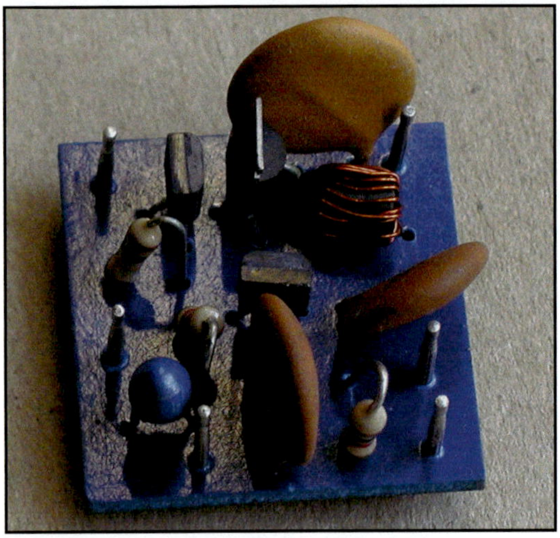

ONER Transmitter Board, which contains only oscillator and power amplifier circuits.

Calling Frequency (see earlier) is a good idea but avoid the frequencies where the 'big-gun' DXers of the amateur radio world operate. I often use the 'listen and pounce' method. I leave the station ready to operate with the receiver on the frequency of the transmitter and do something else, like a little construction work. When I hear a suitable station calling, or ending a contact, on the receiver, I leave my other job and pounce.

Crystals and filtering

Many of the simpler transmitters only have an oscillator and power amplifier built on the board and do require some additions for use on the air. In addition to a 12V supply and key, they need a crystal and a low-pass filter for the required band. **Fig 14.1** shows the usual arrangement. The oscillator will require a crystal to determine its frequency. An ideal choice of frequency is the QRP calling frequency for the band in use. QRP frequency crystals used to be available from suppliers like Les Jackson G4HJZ's AMTOOLS UK but now are more difficult to source.

The GQRP Club sells crystals to members. The crystal should operate at the actual (fundamental) frequency of the band.

Most basic QRP transmitter boards have a 50-ohm output impedance to match an antenna. But it is unwise and even illegal, to use the transmitter without some form of harmonic filtering. QRP transmitters do radiate less output than conventional transmitters but usually the power amplifier operates in 'Class C' which tends to generate harmonic outputs; outputs at twice, three times, etc of the required frequency output. The usual remedy is to add a low-pass filter to the output circuit. This is a tuned circuit arrangement with a cut-off frequency set a little above the required output frequency. The filter attenuates (reduces) the output above the cut-off frequency and allows signals below this frequency to pass. Usually the low-pass filter will have an input and output impedance of 50 ohms. Later in this book I deal with the design and building of low-pass filters.

I keep a range of plug-in low-pass filters for the HF amateur bands. This enables me to plug in a filter and crystal on any small QRP transmitter board. I can use the same low-pass filter board for a variety of applications.

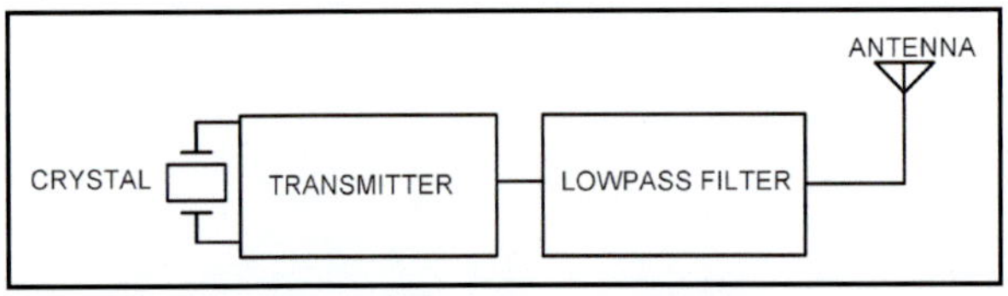

Fig 14.1: A simple QRP transmitter set up

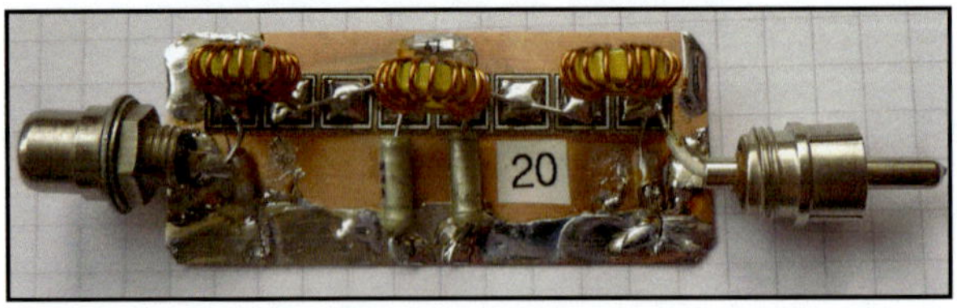

A plug-in low-pass filter for the 20m band

Transmit-receive changeover

The operator will need some method of switching between the transmit and the receive cycles during operation. This will involve switching the antenna between the receiver and the transmitter, and probably some switching of the power supply to enable the transmitter to turn on and off.

Transmit-receive changeover is sometimes called 'QSK', although it is only technically QSK if the changeover between transmit and receive is almost instantaneous. QSK is the Q-Code abbreviation for "I can hear you between my signals" indicating a changeover system so rapid that other stations can be heard virtually at the same as

transmissions are being made. This is often called full break-in operation. To enable this method of changeover with a basic QRP transmitter would involve electronics more complex than the transmitter itself! Most changeover systems for a basic QRP station are very simple.

Fig 14.2 shows a simple transmit- receive changeover system. It uses a double-pole changeover switch; a switch with two sections, operated by the same lever, which changeover between two contacts. One section switches over the antenna between the receiver input and the transmitter output. The other section switches the 12V supply between 12V[R] (12 volts on receive) and 12V[T] (12 volts on transmit). Only the positive side of the supply is switched, the negative side of the supply goes to ground.

The switched 12V is required to enable the transmitter to be used in conjunction with the receiver. The usual arrangement in a simple QRP transmitter is for the key to switch the power amplifier on and off; the oscillator is left running. The problem with this arrangement is that the oscillator will be heard on the receiver when listening to the other stations. The switch provides power to the oscillator stage only during the transmit cycle of operation.

In a crystal-controlled oscillator, switching the power on and off should not cause stability problems. When it receives power, the oscillator locks very quickly on to the crystal frequency. It is a problem for a variable frequency oscillator (VFO) controlled transmitter because a VFO will be subject to warm-up frequency drift. So instead of switching, VFO-controlled transmitters often incorporate some form of frequency off-set circuit to move the oscillator frequency out of the range of the receiver during the receive periods.

The switch is also used for 'netting' the transmitter to the receiver. Through the demise of the separate transmitter and receiver approach in amateur radio, 'netting' has almost become a lost technique. It means locating the frequency, on the receiver, where the transmitter is transmitting. For crystal-controlled transmitter, this is a simple operation. The switch is placed in the transmit position and the receiver is tuned until the oscillator signal is heard. This is where the transmitter signals occur and is the frequency to listen for other stations to contact. In a VFO-controlled transmitter, 'netting' is an altogether more complex process. When a suitable station is located on the receiver, the VFO is switched on and the VFO frequency is moved until the VFO signal 'lies over' the signal from the other station.

Transmit-receive switches are anathema to designers of modern amateur radio equipment. The modern operator expects to be able to push the Morse key or microphone button to change from receive to transmit. Some QRP operators use simple versions of this idea by adding a changeover unit to their transmitter and receiver.

Fig 14.3 shows a simple station where a changeover system has been added. An exam-

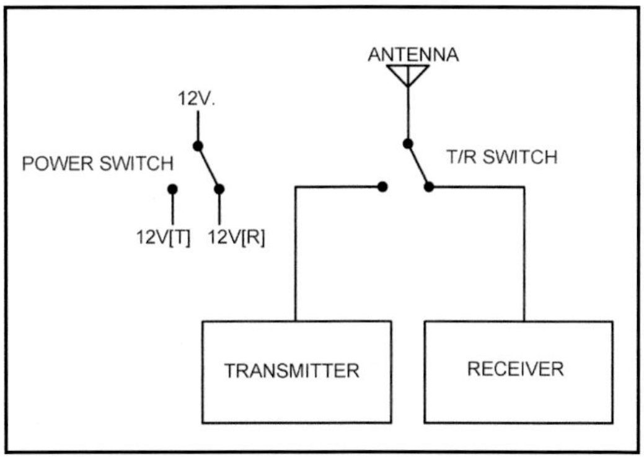

Fig 14.2: A simple transmit-receive changeover arrangement

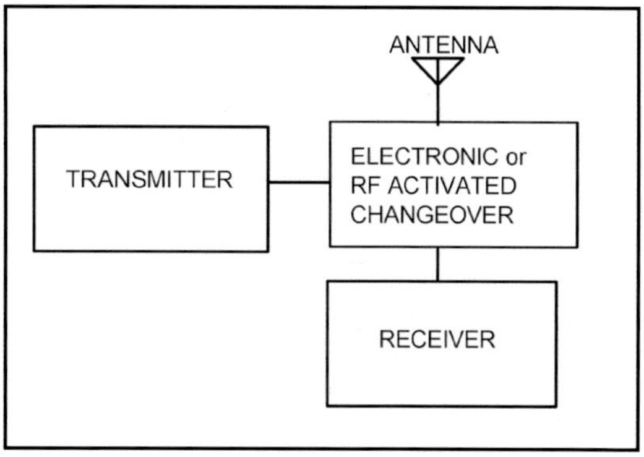

Fig 14.3: Simple automatic transmit-receive switching

ple can be found in the practical projects section of this book. These systems tend to come in two 'flavours': electronically controlled or activated by a radio frequency (RF) signal.

The electronic changeover circuits are controlled by a Morse key. When the key is pressed a relay switches over the antenna and provides the required 12V supply for the oscillator. The relay circuit usually has a built-in delay so that the relay remains switched in the gaps between the sending of Morse code characters. This prevents the relay constantly clattering in and out during the keying operation. Also, constantly switching over the

antenna and switching the oscillator on and off would produce a badly keyed signal. A pause in operating the key at normal speeds will allow the relay to return to the receive position. The circuit may also include an electronic switch to produce a 12V source that follows the action of the keying but without any delay. This may be used as an alternative way of keying the transmitter or activating an electronic muting circuit within the receiver.

Fig 14.4 shows a very basic electronic change- over system, which I have used from time to time. The transmitter is connected directly to the antenna through a screened cable.

The receiver is also connected to the transmitter output via a small value capacitor. This allows the antenna signals to reach the receiver. The value of the capacitor is chosen to be the smallest value that will still allow adequate signals to be heard.

The capacitor reduces all the transmitter output that is fed to the receiver input. There will still be a considerable amount of transmitted signal entering the receiver; so, two back-to-back diodes (D1 and D2) are placed between the input of the receiver and ground. These are any common silicon high-speed switching diodes and they limit the amount of transmitted signal that can reach the receiver.

An RF activated changeover system uses the radio frequency signal of the transmitter to perform the changeover switching. When the RF signal from the transmitter being keyed arrives, the circuit senses and activates a relay which performs the changeover functions. I quite often use this method. In fact, I have an RF changeover unit attached to the back of my favourite station standby receiver, a vintage Davco DR-30.

The photograph shows the four connections to the RF changeover unit. There is a 12V inlet to power the circuit. The box has three screened cable sockets. The receiver input and the antenna are permanently connected to the unit. When I want to use, or test, a little QRP transmitter board, I sim- ply take its output to the 'transmitter' socket, and transmit. The transmitted signal activates the unit and the changeover functions are performed.

Receiver muting

So far, I have not mentioned what happens with the receiver during the transmit cycle. The antenna is removed but there will still be a lot of stray radio frequency signal entering the receiver. This will produce signals far in excess of the received station signals and they will probably sound very 'thumpy'.

The simplest solution is to turn down the receiver volume during the transmit periods. It may sound crude but it can even be helpful. By turning down the receiver audio output to a reasonable level, the transmissions can be monitored as the transmitter is keyed. Most people like an audible indication of the keyed signal, as keying 'deaf' can be difficult.

Another choice is to switch the receiver audio off during transmission. Switching off the whole receiver is a bad idea as it will tend to drift in

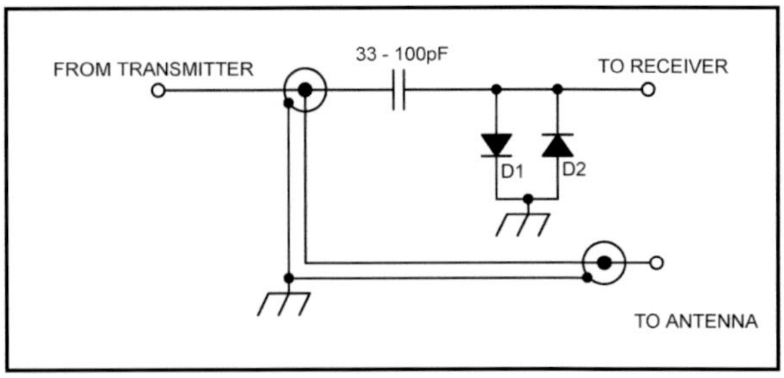

Fig 14.4: Example of a very basic electronic transmit-receive changeover system

An RF activated changeover unit built into a tobacco tin

frequency when first switched back on. There are several ways to mute the audio, but all of them involve modifications to the receiver. The simplest way is to short the volume control potentiometer to ground during transmissions. A more sophisticated method is to build electronic muting into the receiver. The most common electronic method is to insert an FET muting switch in the audio path of the receiver.

Matching the antenna

Assuming that the little transmitter and the low- pass filter have an output impedance of 50 ohms, this may have to be matched to the available antenna.

Many commercial and some homemade antennas such as the half-wave dipole, do match a 50 ohms output, but a lot of amateur radio stations use wire antennas that will require matching.

Fig 14.5 shows the arrangement for matching a transmitter to an antenna. If the antenna does not match the 50 ohms of the transmitter, an Antenna Tuner Unit (ATU) will be required. This is a com- bination of inductance and capacitance contrived so that the transmitter 'sees' a 50-ohm match. Most ATUs have variable controls to allow them to be used over a range of frequencies and with a variety of antennas.

The adjustment of the ATU is checked using an SWR (Standing Wave Ratio) meter. When a mis- match is present some of the signal will be reflected back along the antenna feeder line. The SWR meter detects and indicates the forward voltage and the reflected voltage in the feeder. Because the ATU is doing the matching operation, the SWR meter is inserted between the transmitter and the ATU. The SWR meter sees the impedance presented by the ATU to the transmitter, rather than the impedance of the antenna.

Transceiver operation

All of the above involve the use of separate transmitters and receivers. If a QRP transceiver is being used, the set-up is more conventional.

Fig 14.6: shows the usual set-up for a QRP station using a transceiver. This is drawn to show how all the elements of the station inter-relate. The transceiver will contain its own transmit-receive changeover system. It will probably require connection to the antenna via an SWR meter and ATU to match the station antenna.

Depending upon the mode of operation, the rest of the station could be called the 'operator-radio interface'. The CW station will require a key, and the SSB station will require a microphone. A straight (up and down switch) Morse key is connected directly to the transceiver. If an electronic keyer is to be used,

keying paddles are connected to the keyer, which connects to the transceiver. This is the typical appearance of any amateur radio HF bands station.

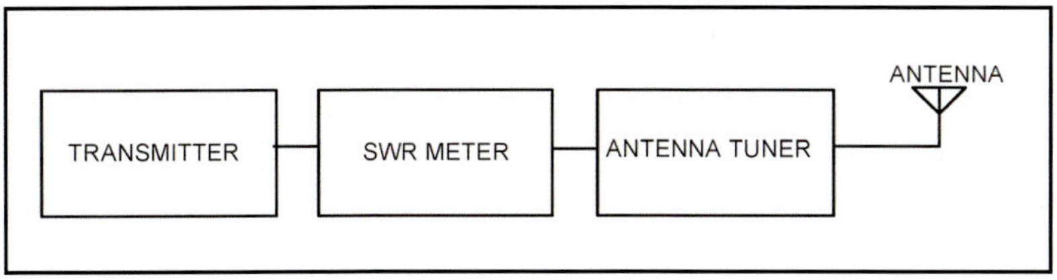

Fig 14.5: Adding an antenna tuner unit (ATU) to the QRP station

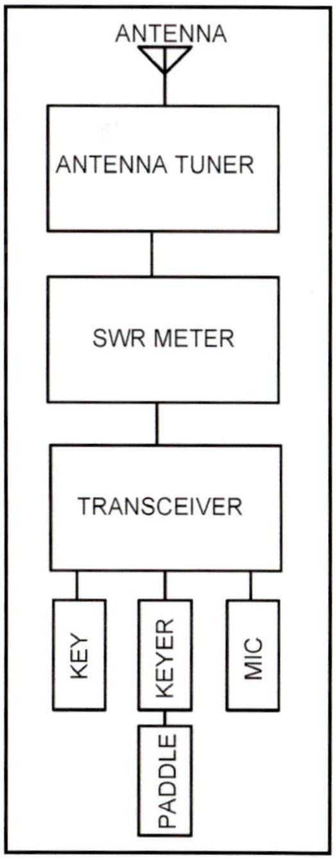

Fig 14.6: A QRP station using a QRP transceiver

The photograph shows the station I used in my holiday lodge in Wales. It is based on a Yaesu FT-707 transceiver which has been 'shot in the foot' so it can only run low power. A 100W transceiver may appear an odd choice for a QRP station, but the FT-707 was available at a cheap price. It is also one of the last of the commercial transceivers where I knew what was going on inside the works, and can get at the circuits if anything goes wrong.

The bottom shelf has the power supply and speaker unit. Above this is the FT-707 transceiver, and the matching ATU above that. The top shelf holds a 2m multimode transceiver along with the SWR and wattmeter that is between the FT-707 and the ATU. All big-stuff but arranged in a classic layout for a transceiver based QRP station.

The GW3RJV station

Accessories for the QRP station

From the above it is obvious that the QRP station may need a few accessories to turn it into a convenient-to-operate set-up. Many such accessories lend themselves to home construction either from scratch or from a kit.

Antenna tuners and SWR meters

Both of these are ideal candidates for home construction. There is a variety of easy to build designs for QRP ATUs and simple SWR bridges. One good source is the G QRP Club Antenna Handbook. Most UK kit manufacturers include an ATU in their list of kits.

One such kit ATU is the Walford Electronics Antenna Matching Unit (AMU). It is a small unit usable over the range 1.8 to 30MHz, working with up to 20W of RF output. The AMU includes a resistive matching bridge with a LED as matching and output indicator. The matching section uses the T configuration to suit a wide range of antennas with balanced or unbalanced feeders. It is supplied with the Walford-style printed circuit board material front panel and all the necessary parts.

A very popular ATU with QRP operators in the USA is the ZM-2. It is a small version of a Z-Match, using only one inductor to cover all the HF bands. The Z-Match is the ATU I have used most often in my home station equipment. The kit is produced by EMTECH who also produce the NW series of QRP transceivers mentioned earlier.

Quite a lot of do-it-yourself ideas for similar ATUs have appeared in recent years. One such is the 'Re-Pete Tuner' by Fred Bonavita, W5QLF, published in SPRAT. The Re-Pete Tuner used inexpensive polyvaricon variable capacitors and a single toroidal inductor to cover all the HF bands. It also incorporated a very simple SWR indicator which, in place of a meter, used an LED.

Automatic antenna tuners are micro- processor-controlled units that automatically select the correct values of capacitance and inductance to match the output of a transmitter or transceiver to an antenna. For some time, I have used the LDG QRP Tuner with my home station. The LDG Z11 Automatic Tuner now replaces this, and is available in the UK from Nevada Radio.

Operation is simple. The tuner is given an output from the transmitter and the operator presses a 'Tune' button. A series of relays, driven by a microprocessor, then select the correct value to match the transmitter output to the antenna. The Z11 can match an antenna in the impedance range 6 to 800 ohms over a frequency range of 1.8 to 30MHz. The latching relays are powered off once tuning has taken place and the microprocessor also features a 'sleep mode' to turn the Z11 off when not in use.

SWR Meters and indicators are very simple to make and I will describe some ideas later in the book. My favourite SWR tune-up device is the 'Stockton In-line Wattmeter' from David Stockton, GM4ZNQ, which has become something of a legend with QRP constructors. Although technically not an SWR meter, it does read forward and reflected (reverse) power in watts. I have built several of these and found them to be valuable assets in my QRP station. Apart from the usefulness of indicating forward and reflected power, the circuit remains in-line with the output of a transmitter or transceiver, continuously reporting the power being transmitted.

For many radio amateurs, QRP is synonymous with radio construction. A casual glance at the range of journals produced by QRP groups will show a variety of practical projects to build, from the simplest small accessory to complex state of the art transceivers. The emphasis is somewhat different on either side of the Atlantic. In the USA, kit building is the commonest form of QRP home construction but in Europe many more people build QRP projects by gathering the designs and parts for themselves.

The variety is amazing. Some QRP operators run completely home built stations, but these are in the minority. Most QRPers probably have a main station which is commercial or kit built, and experiment with home-built projects. That is the way I work. My short experience of a completely home built station was marred by the problem of wanting to constantly modify and update my homemade projects, I rarely had a complete station to use at any one time!

Building equipment obviously enhances that old amateur radio ideal: "self-training in radio communication". Certainly, many radio constructors relish the educational aspect of their hobby but for most people it is simply fun. Some build complex projects but for many their favourite type of project would be something which costs very little in parts, can be built in an evening or two and gives the satisfaction of producing something with one's own hands. QRP lends itself to such an approach in the practical side of the hobby.

Inside the Re-Pete Z-match ATU by W5QLF

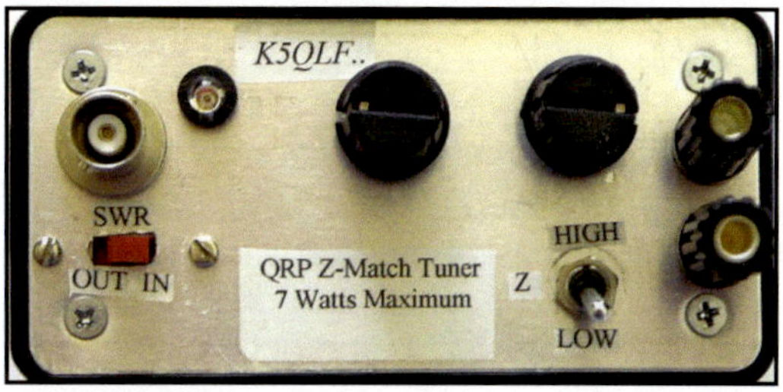

The outside the Re-Pete Z-match ATU by W5QLF

The Walford AMU (a kit antenna tuner)

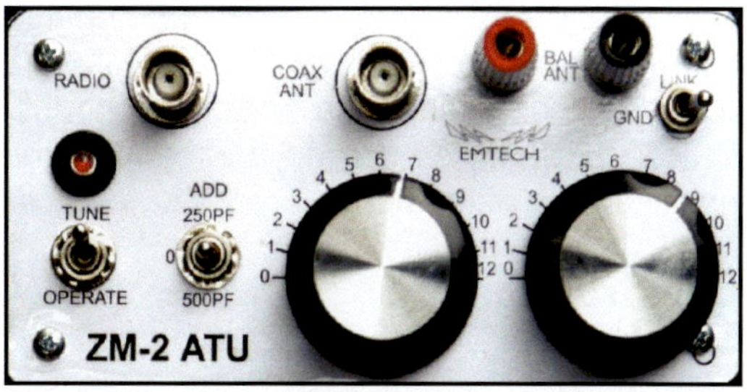

The ZM-2, a kit version of the Z- match ATU from EMTECH in the USA

Chapter 15

Home-made QRP

For many radio amateurs, QRP is synonymous with radio construction. A casual glance at the range of journals produced by QRP groups will show a variety of practical projects to build, from the simplest small accessory to complex state of the art transceivers. The emphasis is somewhat different on either side of the Atlantic. In the USA, kit building is the commonest form of QRP home construction but in Europe many more people build QRP projects by gathering the designs and parts for themselves.

The variety is amazing. Some QRP operators run completely home built stations, but these are in the minority. Most QRPers probably have a main station which is commercial or kit built, and experiment with home-built projects. That is the way I work. My short experience of a completely home built station was marred by the problem of wanting to constantly modify and update my homemade projects, I rarely had a complete station to use at any one time!

Building equipment obviously enhances that old amateur radio ideal: "self-training in radio communication". Certainly, many radio constructors relish the educational aspect of their hobby but for most people it is simply fun. Some build complex projects but for many their favourite type of project would be something which costs very little in parts, can be built in an evening or two and gives the satisfaction of producing something with one's own hands. QRP lends itself to such an approach in the practical side of the hobby.

Construction fun

One of the delights of QRP construction can be to build something just for the fun of it. Rick Weber, W9QZ, is a fine example of a fun constructor. Joining in an internet constructor's group called 'Junk Box Wars' Rick turned his hand to making a

transmitter completely from scrap items. No good rea- son for doing this apart from simple amateur radio fun. I might even suggest that this is what all hobbies are about! In Rick's own words: "I set up my own personal challenge to build a QRP CW transmitter for 80m that used only one commercial electronic component, a vacuum tube (valve) and only non- electronic junk commonly found around the house. "I started with a Hartley oscillator design to use one very old 27 triode vacuum tube made in the late 1920s. No commercial resistors, capacitors, chokes, or variable tuners were used. The capacitors were made from Diet Coke cans and clear packing tape; two .002uF and one 250pF.

W9QZ's Coke transmitter was made from scrap. I made the 500pF variable condenser from one diet coke telescoping over another

The 616 Valve transmitter built by SM7UCZ

The G4RAW TAG transmitter (the G3RJV version)

one with packing tape insulation. The RF choke is 160 turns of wire on a ball point pen body. The 10k grid resistor was made using the old science fair trick of a soft graphite pencil rubbed on cardboard. Two paper clips provided the resistor leads. Twelve turns of wire on a plastic pill bottle for the tank coil. The crazy thing worked! My first QSO was with Bob Howard, K0RDF, about 350 miles away. My RST report was 239. The best I could tell, this thing was putting a little over a watt to the antenna. Why did I use a tube (valve)? Well the best radios glow in the dark!" Another QRPer who builds valve equipment purely for the fun of it is Johnny Apell, SM7UCZ.

Each year, at Christmas time, Johnny builds himself a QRP project purely for fun. Very often this is a QRP transmitter built in the 1920s breadboard style.

The transmitter in the photograph is one of the SM7UCZ 1920s style projects, which he was kind enough to give to me on one of his visits to England. It is a beautiful piece of construction which also works and puts out a few watts of RF on the 80m band. There are many other QRP constructors who enjoy using valves, and there has been quite a revival of interest in older circuits and designs of 70 or 80 years ago.

Fun in simplicity

The one night or one weekend project can be very satisfying. Few people these days appear to have jobs where at the end of the day they can say: "I made that". It is a gratifying pursuit to be able to make something from start to finish that works and can be used in practical radio communication. The GQRP Club's SPRAT often contains ideas for such projects and they are popular with the members. One such project was the TAG Transmitter by Steve Ortmayer, G4RAW (SK), so called because the whole transmitter is built on one small 'tag board'.

The transmitter is a version of the 'Universal Transmitter' by Wes Hayward, W7ZOI. The photograph shows my version, which I built to work on the 40m band. The transmitter includes a low-pass filter and just needs a crystal for the band and an Altoids mint tin, a favourite with QRPers, houses a receiver built 'ugly-style', a 12V supply. I built mine in about an hour and went straight on the air on 40m, contacting five European countries in the same evening. Beginning with my existing station receiver and adding a simple board, I had made myself a basic amateur radio station in just over an hour. Not only that, it was capable of viable amateur radio

The G4RAW TAG transmitter (the G3RJV version)

An Altoids mint tin, a favourite with QRPers, houses a receiver built 'ugly-style'

communication, albeit on only one band.

For some constructors simple can equal small and they take great delight in building small projects, very often in little tins. A recent craze amongst QRP constructors has been to build their projects in Altoids Mint tins. It is a curious idea that so many radio constructors in the USA have opted to build their little projects in a "curiously strong peppermints" tin made in Great Britain! In fact, some projects, like the Kanga Products FOXX-3 transceiver have been designed to fit in an Altoids tin. This has been a source of great fun and probably an asset to mint manufacturers!

The receiver in the photograph is an example of minimalist construction. It is a direct conversion receiver for 40m built 'ugly-style', without a circuit board, and components wired point to point. It is as easy as choosing a circuit, in this case a version of the Sudden receiver, gathering the parts and a container and then wiring them together to complete the project. For those who enjoy a practical hobby, a project like this is an evening of low-cost fun with a practical outcome.

We have seen the ONER transmitter; a little transmitter board measuring one-inch square. KPS Kang, VU2KLA(SK), of the Indian QRP Club produced a club printed circuit board that fitted a transmitter on to an even smaller board . . . just! The VU-Transmitter measured 0.8 x 0.8in (20 x 20mm) square and held a basic crystal oscillator feeding a Class C power amplifier. The VU Club were kind enough to send me a printed circuit board, and the photograph shows my VU-Transmitter.

Perhaps it is slightly cheating because some components overlay others but it does work. Using a 2N2222A transistor feeding a 2N3866 amplifier my board gave an RF output of a little over half a watt. I even had contacts with several European countries using the diminutive transmitter.

All of the above is not to say that the only interest within the QRP building fraternity is in building simple equipment. A good many constructors build very sophisticated projects. There are a fair number of high-tech home -built QRP stations about. Earlier we saw the fine GM3OXX (SK) home made all band transceiver which has achieved such excellent results with one watt on the HF bands.

Jan Verduyn, G0BBL, together with Alan Rowe, M0PUB, and John Law G8BTR, were

The tiny VU-Transmitter board

Jan Verduyn, G0BBL, showing off the QRP2004 Project

the design team behind the QRP2004 Project.

The QRP2004 is a multi-band HF receiver for the home constructor, which can be expanded to a complete transceiver. It is a development of their QRP2000 Project; a phasing direct con- version radio based on the KK7B R2/T2 modules, using a synthesised VFO and covering 1.8-30MHz. The design and practical implementation of the project can be followed on a website set up for that purpose. Details of the project can be found at: www.yumpu.com/en/document/read/25615813/the-qrp2004-all-band-hf-transceiver.

The QRP2004 is a fine example of what can be done when a group of enthusiasts decide to share their skills. Each of them brought their own talents to the project and they are keen to share their efforts by publishing the information freely on a website.

Another example of an interesting and unusual QRP project is a Polyphase, Tayloe-Detector, Receiver built by Hans Summers, G0UPL. Hans' receiver (see: www.hanssummers.com/polyphase.html). The heart of the receiver is a FST3253 Tayloe switching mixer with a Franklin VFO using a 74HCT04 inverter chip, stabilised by a magnetically coupled Huff and Puff circuit and followed by an 8-column polyphase audio network. The whole project is built using an ugly-style, 'build a little – test a little' technique. The receiver shows how the QRP constructor can develop a project using interesting techniques, and experiment and modify as the project grows.

A QRP construction story

What follows is the story of how a simple little circuit idea inspired a whole range of QRP constructors and experimenters to take a basic idea and work out their own ideas and versions. The story has almost taken on the proportions of a saga!

In the summer of 1983, SPRAT published a design by George Burt, GM3OXX (SK), titled 'The FOXX Mini-transceiver – a transceiver on a 2" x 2" printed circuit board'.

The circuit was novel in that the same bi-polar transistor served as the power amplifier for the transmitter and the mixer for a direct conversion receiver. (The FOXX design has its own history. In 1997 I wrote an article for SPRAT, 'The FOXX 2 – an old favourite revisited' in which I described some updates for the original GM3OXX circuit.

This was quickly followed by the FOXX-3 a further upgrade by Derek Alexander, G4GVM. This circuit became the basis for the popular Kanga Products kit; a complete transceiver in a mint tin!)

In 1992, I received a letter from a Russian QRP operator Oleg Borodin, RV3GM, enclosing a circuit idea for SPRAT. It was called the 'Micro-80, the simplest and smallest transceiver'. The circuit was published in the summer issue and spawned a whole series of follow- up ideas.

The Micro-80 certainly is a

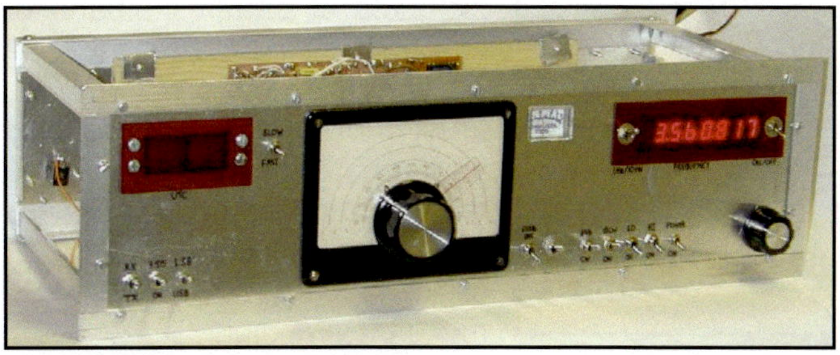

The G0UPL Polyphase receiver

The FOXX-3 a complete transceiver in a mint tin

simple circuit and follows the GM3OXX idea of using the power amplifier as the receiver mixer. For those who understand simple circuits, let us follow it through.

Transistor T1 is a crystal oscillator using a version of the Colpitts design. The two capacitors, C1 and C2, form a capacitive tap to allow feedback between the emitter resistor, R2, and the base of the transistor. The oscillator output is coupled via C3 to the power amplifier transistor, T4. When the key is closed, the signal is amplified by T4 and appears at the tuned circuit formed by L and C4, which resonate at the required output frequency. The output goes via C* (valued selected in use) to the antenna, hopefully through some extra har- monic filtering to comply with emission regulations. T1 and T4 form a basic transmitter with an output, depending upon the transistors in use, in the order of half a watt.

Consider what happens when the key is open. The oscillator, T1, is still running and still feeding a signal to the base of T4. The collector of T4 is joined to the tuned circuit, which is connected to the antenna. A bi-polar transistor could be thought of as two diodes, back to back. T4 is receiving two signals; the oscillator and the

tuned input from the antenna, and will function as a simple mixer.

In a direct conversion receiver, a mixer which has inputs from radio signals at a desired frequency and an oscillator signal at the same frequency will produce audio products at either sideband of that frequency. Because the local oscillator signal is present all the time, the method is only really appropriate for receiving Morse or single-sideband (SSB) signals. Both of these modes require a local oscillator to add a carrier signal to make them audible.

When the key is open, T4 will

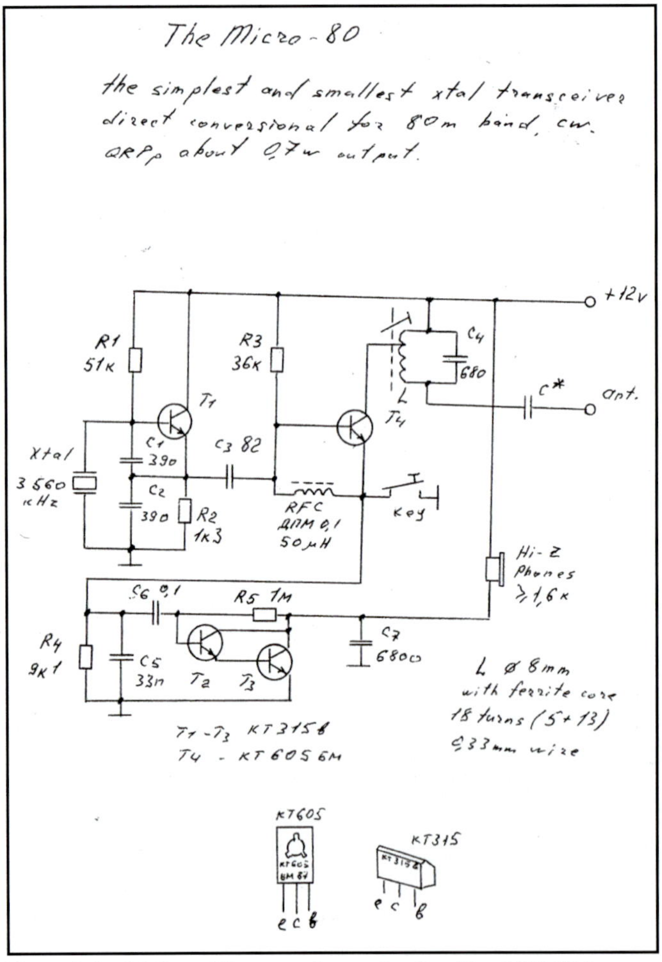

The original Micro-80 transceiver submitted by RV3GM

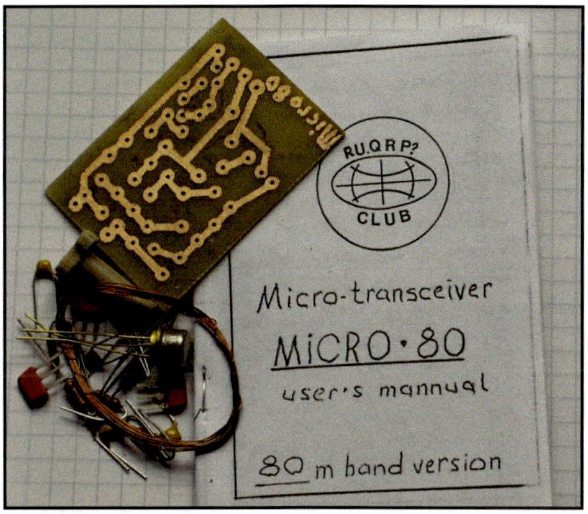

The RV3GM Micro-80 kit

act as a mixer and the audio product will appear at the emitter. In the Micro-80, the emitter of T4 is connected to a small two-stage audio amplifier, T2 and T3. T2 and T3 are arranged as a Darlington pair to give enough audio signal to drive high impedance head-phones. The Micro-80 is not only a simple transceiver but it operates with complete 'break-in' operation. When the key is down it transmits and when the key is up it receives!

RV3GM even produced a kit version of the Micro-80 which was available through Rex Harper W1REX (manual here: https://docplayer.net/49931479-Micro-80-d-minimalist-80m-transceiver-original-design-by-oleg-borodin-rv3gm-updated-and-kitted-by-rex-harper-w1rex.html)

Obviously, such a simple circuit has some drawbacks. The transmitter output is low and really does require extra harmonic filtering after the power amplifier tuned circuit. The receiver is very basic. All of the receiver gain comes from the audio amplifier and this is only enough to hear strong signals, although these are probably the ones which would be contacted with such low power. The chief problem is the lack of selectivity; the ability to sort out the required signals in favour of the unwanted signals. The only frequency selective element is one tuned circuit (L and C4), and unwanted signals, especially strong broadcast signals, are likely to break through.

But having said all that – it is a complete radio station, it is simple and easy to make and it can be fun to use. It certainly inspired other circuits.

The DARA Hamvention, held each year near Dayton, Ohio, is the largest amateur radio event in the US. For many years the event has been a meeting place for QRP enthusiasts who book their own hotel and host many special QRP events. At the 1995 Dayton Hamvention, the Northern California QRP Club (NorCal) sponsored a building contest based on the PIXIE-2. The PIXIE is the Californian version of the Micro-80 designed by Dave Joseph, WA6BOY, from RV3GM's idea.

In the PIXIE, a low-pass filter replaces the single tuned circuit and the audio amplifier became an LM386 audio amplifier chip. Both of these changes offered considerable improvement of the circuit. Very quickly the PIXIE-2 was updated by Derrick Webber, G3LHJ, to give frequency off-set between transmit and receive. Then the doyen of home construction, Doug DeMaw, W1FB (SK), submitted his version of the circuit with offset and audio filtering. Versions of the Pixie proliferate to this day.

Six have been published in SPRAT. QRP constructors seem to want to take this elegant little circuit and personalise it. Johnny Apell, SM7UCZ, produced an attractive printed circuit board for his 'Älva' transceiver. An Älva is the nearest Swedish mythical figure to a

The Älva, the Swedish Pixie

Pixie. The Älva has an extra stage of audio amplification before the LM386 audio chip.

The Älva circuit diagram gives an idea of the difference between the various Pixie clones and the original Micro-80 circuit. The early problems of selectivity and gain are aided by the addition of better input and output filtering and increased audio amplification. The transceiver is still rudimentary but novel and fun to build and use. It is an ideal beginner's project to build rather than an ideal beginner's transceiver to use.

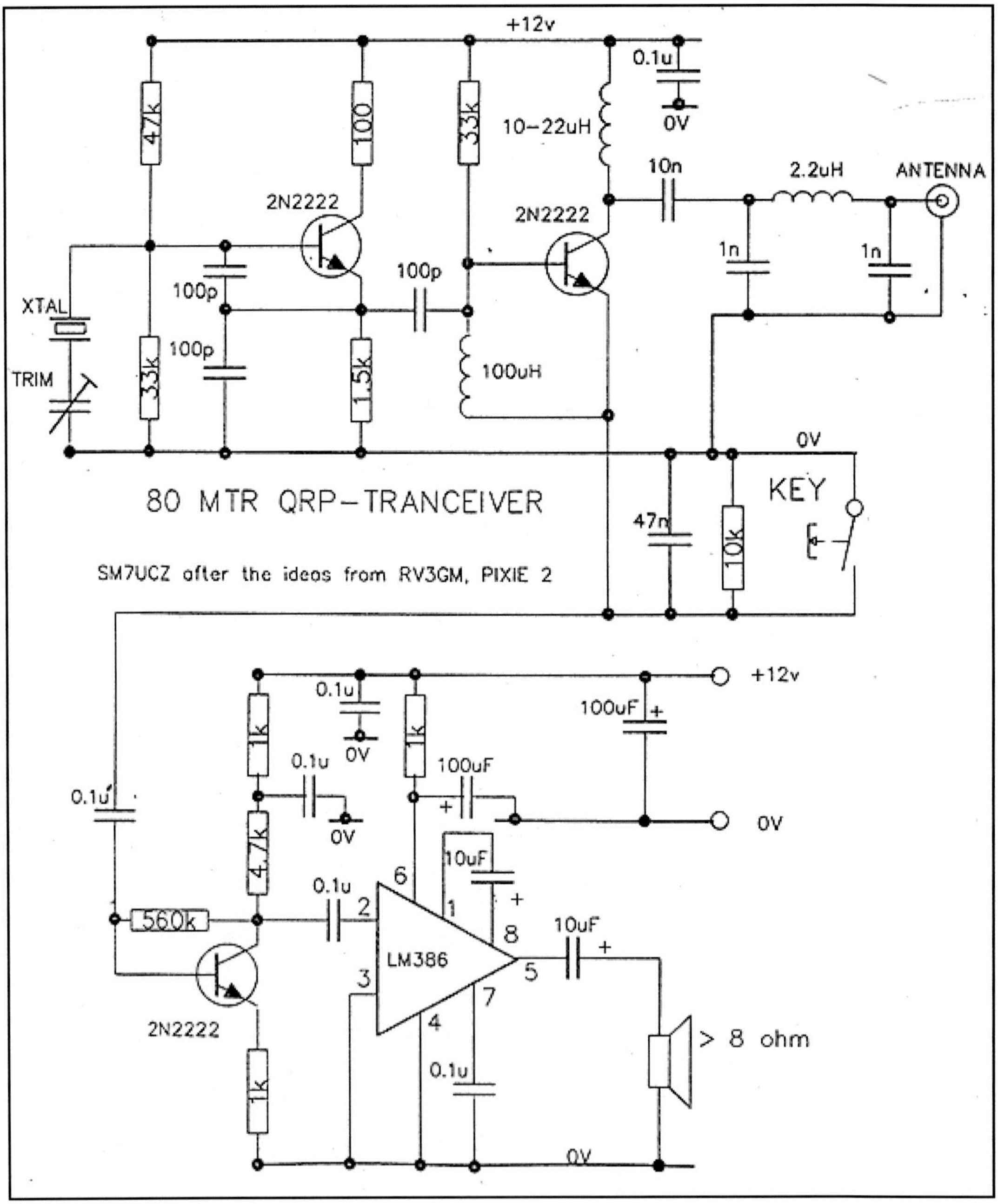

The Älva transceiver circuit diagram

So here we have an idea that has run for nearly 30 years through the enthusiasm of QRP radio builders. Not exactly the cutting edge of technology but a source of enjoyment for many radio amateurs for a long time.

The Buildathon

In recent years many people have been introduced to amateur radio construction through taking part in a 'Buildathon'. A buildathon is a group radio construction event where novice builders build a project under the guidance of more experienced constructors. All the participants build the same project and are supplied with all the parts and instructions. The idea originally came from the American QRP ARCI group who hold an event called Four Days in May as part of the Dayton Hamvention, the US's largest amateur radio convention. Several years ago, they introduced an annual buildathon to Four Days in May, since when the idea has been copied by many other groups.

In the UK, Buildathons were made popular by a group of radio amateurs in Bath led by Steve Hartley, G0FUW. The Bath team produced a video describing the buildathon and giving advice on how to run local buildathon events. The Bath team have also run Buildathons for other groups including at the GQRP Club convention. If you are new to radio construction look out for local buildathon events. It is a good way to begin.

A Buildathon workbench at the GQRP Club Convention. Everyone can join in, irrespective of age or gender

Marital harmony! Me assisting my wife Jo-Anna, G0OWH, with her buildathon project

Chapter 16

The QRP Workshop

Some people are attracted to the QRP aspect of amateur radio because of the opportunities for building their own equipment. Most QRPers have some kind of workshop for their building and experimentation. For those without existing workshop facilities this can sound daunting but the QRP workshop is simple to set up.

I have heard it said that the days of practical amateur radio construction have gone but it has never been so easy to be a radio constructor. When I began construction, I spent many hours in a cold gar- den shed working on making large metal chassis and mounting valves and big components. These days the QRP workshop can be compact. The physically smaller scale of amateur radio means that a workbench could be a wooden tea tray with a soldering iron and a few hand tools, which can be cleared away after use. For many years my 'construction bench' was such a tea tray, bought at a jumble sale, that could be brought out and placed on any convenient table. Later my 'construction corner' became one end of a large bench that also contained the radio station and a lot of untidy clutter. My working space was only about 2 feet wide by 18 inches deep (650 x 490mm).

I have also heard it said that construction was easier in the days of government surplus and radio junk shops. In real terms, electronic components have never been cheaper. In the heyday of the government surplus shops, I recall saving up to buy parts to build my radio equipment. The cost relative to average income these days is such that buying the parts is rarely a problem. Sourcing the parts can be a more difficult so I have included a list of component traders later in this book. An increasing problem is the move within the electronics industry away from components with leads and towards surface mounted devices.

Another common complaint is the lack of buildable radio projects, as amateur radio magazines now appear to publish fewer practical articles. There is no such problem for the QRP equipment builder. In addition to a fine collection of books on QRP which include practical projects, there are several QRP clubs

which publish small magazines, many of them with excellent practical articles. This is also the age of the kit builder. There have never been so many amateur radio kits, most of them aimed at the QRP enthusiast.

Equipping a workshop

What is needed in an amateur radio workshop depends upon the scope of the work. For some people this will only be what is required to build a radio kit. The basic workshop for building boards is quite modest; a few hand tools, a means of soldering and some rudimentary test equipment. The first piece of advice is to buy good quality items. It is a mistake to buy cheap tools; good tools will do a better job and probably last for a lifetime.

Tools

In preparing this section I went over to my workbench and looked at what tools were lying around, probably the ones I use most often. Although I have quite a large toolbox under the bench many of the tools spend most of their life in the box. On my bench I had a few screwdrivers, a small pair of pointed pliers, side cutters and diagonal cutters, a small knife, and an old suede shoe cleaning brush that I

Some hand working tools

use for cleaning circuit boards.

Other tools will be required for cutting and shaping materials, and dealing with the hardware side of a project. Some of these may be part of an existing household toolbox and some may have to be bought according to need. Those who choose to make their own printed circuit boards will require additional specialist equipment.

As with hand tools so with soldering equipment, buy good quality items. Avoid the cheap soldering irons sold in utility stores and look for the sort that might be found on a professional workbench. For many years the soldering iron of my choice has been a Weller Temperature Controlled Soldering Station. They have a selection of easy to change bits that offer choices in soldering temperature. The bits rated at 700 degrees are suitable for most amateur radio construction applications. A non-temperature-controlled iron increases its bit temperature the longer it remains plugged in.

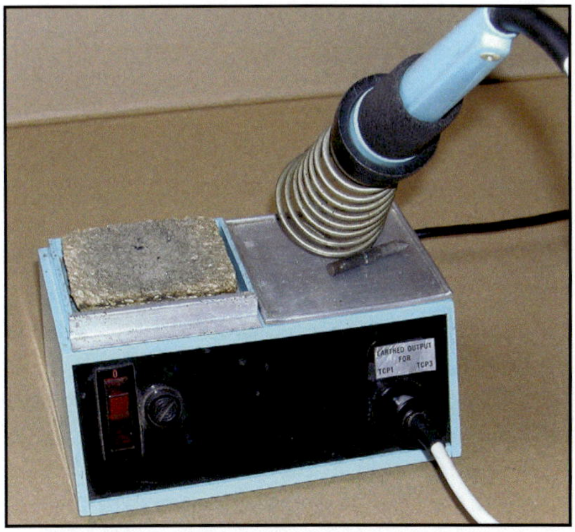

A temperature-controlled soldering station

Soldering stations are quite expensive. If the reader does not wish to begin with a temperature controlled soldering station, the best alternative is to buy a soldering iron from a reputable manufacturer. My stand-by iron is an Antex 25 watt soldering iron with a 1/8th inch (approx. 3mm) bit. This will require the addition of a soldering iron stand. Trying to use a solder iron without a stand is a recipe for burns, on tables and carpets as well as fingers.

My soldering station has a built-in sponge to wipe the bit of the soldering iron. It is important to keep the bit clean.

If a cleaning sponge is not an integrated part of the soldering equipment, obtain a soldering sponge or keep a small damp cloth at hand to clean the bit.

In recent years I have used a Soldering Iron Tip Cleaner with low-abrasive brass shavings in a metal cup. To clean the tip, simply push the hot tip into the shavings a couple of times

Soldering iron tip cleaner which uses brass shavings

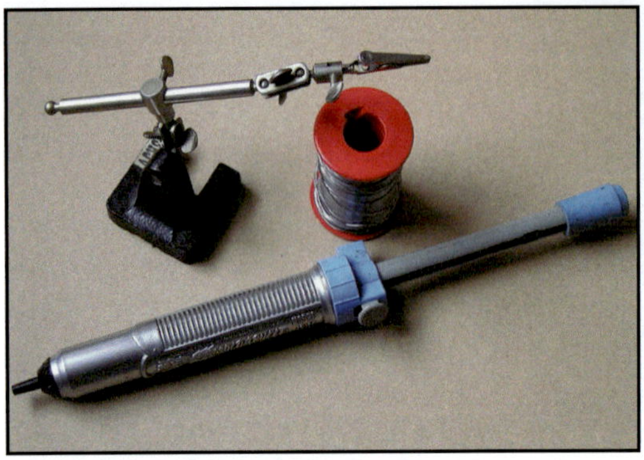

Wire solder and a de-soldering pump together with a 'Third Hand'

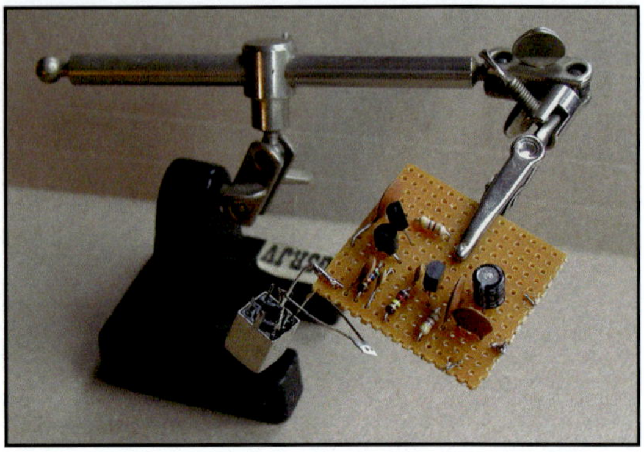

'Third Hand' holding a circuit board. This device sometimes comes with a built-in magnifying glass

for a shiny bright tip ready for use. There is no need for wiping on all sides as with a sponge as the tip emerges far cleaner than with a sponge and at full temperature. The brass shaving pad should last for years, but replacement pads are available.

Use solder wire with a core containing flux to enable a smooth flow of molten solder. It is best to avoid un-branded solder wire that may contain flux with a high acid content. The 'Multicore' brand of soldering wire has five cores of non-corrosive flux and I have used it for many years with good results. The 60/40 (60% tin, 40% lead) alloy is good for general purpose and small component wiring. Health and safety concerns have also led to the introduction of lead-free solder, which is mainly tin and needs a hotter soldering iron to melt it.

All soldering gives off fumes and is best done in a well-ventilated area. Commercial workshops would probably have a fan to remove fumes. This may seem like over-kill for the amateur but it is easy to install a small blower fan, of the type used in equipment and computers, near to the soldering position. It does make working with solder a much more pleasant experience.

Two other soldering tools I use are a Third Hand and a de-soldering pump. The Third Hand is a small clamp for holding items, which are being soldered. It has a base, swivel mountings and a crocodile clip to hold the part or board. I find them very useful.

Sometimes a soldered component has to be removed. This is done by melting the connection and pulling out the part. Removing the solder from the joint with a de-soldering pump helps to free the leads. Mine is a large pump, bought in the US, but there are many smaller pumps that do the job well.

Basic test equipment

Sometimes it is assumed that an amateur radio workshop will contain a lot of expensive test equipment. Most simple QRP projects can be built using the minimum of test equipment. There are a few essentials but after that it is a question of obtaining more advanced equipment as construction skills increase and the opportunity arises. Most of my test equipment was either bought second hand or is home built.

The Multimeter

The real essential in a workshop is a multimeter; a meter that can be switched to measure voltage, current and resistance. It may also have other functions. They come in two 'flavours', analogue and digital. The analogue types have a needle indicating the values against a scale and the digital types have a liquid crystal (or similar) display. These days digital meters are the more common. If the workshop is only to have one meter, I would suggest an analogue type. In practice meters are often used to measure circuit changes

An analogue multimeter.

An inexpensive digital multimeter

or adjustments and it is easier to observe a moving needle than watch a series of dancing numbers. Accuracy can be sacrificed by using an analogue meter but I rarely want to know I have 4.473V, most times I just want to know that I have about four and a half volts.

Digital multimeters are now available so cheaply, that I would advise the beginner to obtain a reasonable analogue instrument and also buy one of the cheaper digital types.

Like most things - you get what you pay for! A 20k-ohms/volt analogue meter is very useful, and practical, to have. Adding a digital meter may not be expensive. They are available for under

£10. But for a little more money, some digital meters will offer other useful test facilities. I have one that will test values of resistance and capacitance and some include a basic transistor tester.

Having acquired a multimeter, another little essential for the QRP builder is a diode probe to use with the meter. The diode RF probes enable the constructor to measure radio-frequency (RF) voltages in a circuit. It is possible to buy an RF probe, but any self-respecting constructor will build one as they only contain four parts. The building of an RF probe is shown later in this book. The probe can be made to measure RMS (root mean square) or peak voltages. The use of the RF probe is one example where an analogue meter is better than a digital meter.

Another useful item of test equipment is a short- wave receiver. Any amateur radio station will have one or more receivers, which may have coverage over the whole HF spectrum. A receiver is very useful for checking the output of an oscillator or transmitter. It can be used to listen for a generated signal. Many receivers have a digital readout of frequency and can accurately measure the frequency of a signal. By merely listening to the signal, it is possible to hear how 'pure' the signal is and if it is frequency stable. Do not underestimate the usefulness of a receiver as a piece of test equipment. I even have a 'skeleton' direct conversion receiver, which I drive from a signal generator,

A home-made RF probe is simple but very useful

to listen to oscillators at odd frequencies.

A dummy load is a very useful make-it-yourself item of test equipment for the builder of QRP transmitters. Most transmitters are designed to give an output with an impedance of 50 ohms and most of them do not like being used without an antenna or some other load at that impedance. Testing a QRP transmitter without a resistive load to dissipate the output power can be an easy way to destroy an output transistor.

All that is required is a resistance of around 50 ohms capable of handling the power from the transmitter. In QRP work this is not too difficult because the resistance has only to handle a few watts. Usually the power capability can be achieved by connecting several lower wattage resistors in series or parallel (or combinations of the two) to obtain 50 ohms and share the power handling between the resistors. Wire wound resistors are to be avoided because they introduce inductance to the circuit and can give odd readings. A dummy load may be used with an RF probe and a meter to measure the output of a transmitter. With suitable calculations this will be a wattmeter.

Commercial and home-made dummy loads: The home-made one is five 10-ohm, 1-watt resistors connected in series

As the workshop develops, other useful items of test equipment may include a radio frequency signal source; signal generators are often available as inexpensive second-hand items. A digital frequency counter is useful and some are available in kit form. Many pictures of amateur radio stations show an oscilloscope but often the ones I see pictured are only capable of displaying low frequency signals and are probably more decorative than useful. If buying an oscilloscope ensure the bandwidth enables it to see signals right across the HF spectrum and learn how to use it. Too many oscilloscopes just lie in the workshop looking impressive and are hardly ever used. One approach is to look for multi-use instruments. One of my favourites multi-use items is the MFJ SWR Analyser. Its fundamental use is to check the SWR (standing wave ratio) of an antenna but is so designed that some of the other internal functions are available for use. The MFJ SWR Analyser can also be used as a digital frequency counter and a radio frequency signal source (with digital indication). A small add-on accessory can also enable it to be used as a dip meter to measure the resonant frequency of a tuned circuit. Four test functions in one instrument make this a very useful item of test equipment.

Another multi-purpose item that gets a lot of use on my workbench is the AADE L/C Meter. AADE stands for 'Almost All Digital Electronics', a small American company which closed in 2015 when its founder Neil Hecht passed away. The meter measures inductance and capacitance directly on an LCD display and is capable of measurements over an amazingly

MFJ SWR Analyser

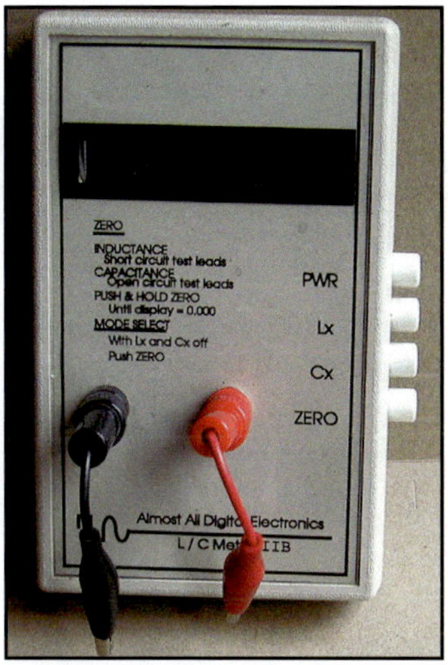

AADE Inductance/Capacitance Meter

Velleman PCSGU250 - USB PC Scope and Generator

wide scale. It is ideal for checking home wound, or unknown, inductors and is good for measuring the capacitance and range of variable capacitors. This meter may still be available second hand.

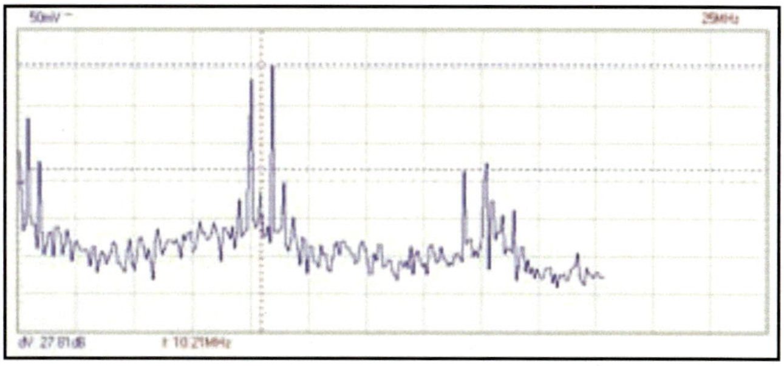

Velleman display of a Double Sideband signal (DSB) with second harmonics of a DSB signal also shown.

Oscilloscopes can be multipurpose in what they can do. One approach is to use a PC-centred oscilloscope such as the Velleman unit. This allows not only waveforms to be displayed, but also stored and transferred to the computer as a permanent record. Another mode of operation is as a spectrum analyser using Fast Fourier Transforms (FFT) to convert from waveforms to spectra. Furthermore, another approach is to have a stand-alone oscilloscope/spectrum analyser which can be connected to a computer and with the ability to download its already stored information to the latter. The Hanmatek and similar designs are available from a range of Far East companies at comparatively reasonable prices. The spectrum analyser can supply very useful information apart from signal spectra, such as distortion and out of band signal information which is very important in today's crowded radio spectrum.

Chapter 17

Construction Tips and Techniques

General tips for the would-be constructor

Tip 1 - Gather all the parts: I never begin work on a construction project before I have all the parts to hand. There is nothing so frustrating as not being able to complete the task for lack of a part or two. Sort the parts out according to type and value, perhaps using the polystyrene tile method described later in this chapter. A guide to recognising resistors and capacitors can be found at the back of this book. Having all the parts available is also essential for some of the methods of construction I will describe.

Tip 2 - Learn to read a circuit diagram: The would-be builder of a QRP project will probably find that the available information is a circuit diagram marked with the type and value of the components required. Some people describe a circuit diagram as being like a map. If you wish to stick with that simile, remember it is more like a London Underground map than an Ordnance Survey map. It simply shows what goes to where but the scale and relative placement are not shown. Look at a basic electronics book to learn the symbols that represent each type of component. Circuit diagrams can be fascinating. I spend many happy hours looking at them and building phantom projects in the air.

Tip 3 - Have faith! Radio construction is oddly like QRP operating; those who think they will succeed, very often do. Beginning a project by believing it will work tends to lead to a successful conclusion.

Tip 4 - Get some clip leads: Anyone looking at my, usually untidy, radio workbench will notice to the right of where I solder, a rather splendid brass toilet roll holder. This holds my collection of clip leads, connectors and test leads. I rarely manage to complete any piece of electronic construction with having to use some clip leads somewhere in the process. Clip leads are short, plastic covered lengths of wire with a crocodile clip at each end, and my collection of leads is in constant use for any temporary connections I wish to make during the process of building and testing a project. My project boards usually see their first application of power via clip leads and I use them to join various boards together during testing. Get a set, or more than one set . . . . *they are so useful.*

Tip 5 - Become good at soldering: Most kits and home built projects that end in failure do so because of poor solder joints. The circuit diagram is probably right; it must have worked for someone to get published. The parts are probably not faulty; new parts rarely are. If something fails check the soldered joints first.

A quick guide to soldering

Unless the simple art of soldering is mastered, electronic construction will be fraught with problems. Good soldering is not difficult if a few simple rules are followed.

Soldering is not gluing

Soldering is a process of amalgamating metals to provide a good electrical connection. Solder is metal alloy with a low melting point designed to flow onto the surface of another metal to create an electrical connection.

Keep things clean

Dirt, grease and oxidation can prevent a good solder joint being made. Wipe leads with fine emery paper or scrape them with a knife until the metal shines. I use a small suede brush to clean the pads on a printed circuit board. The soldering iron bit should be 'tinned' before it is used. Tinning is adding a thin layer of clean solder to the hot tip. Melt some solder on to the tip and wipe it on a wet soldering sponge or damp cloth to give a shiny surface. It is also worthwhile tinning the leads of the component to be soldered by heating the wire and flowing on a layer of solder.

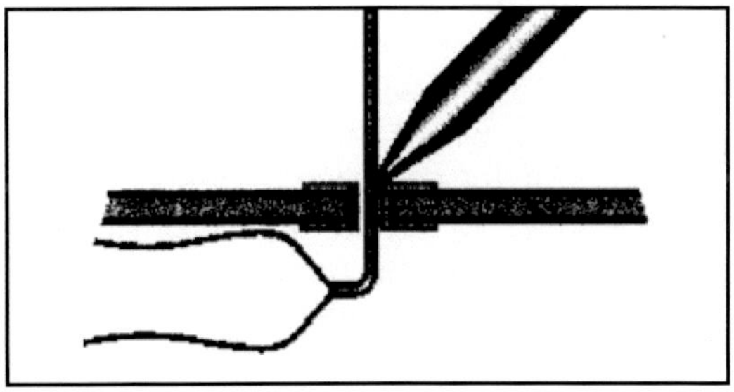

Fig 17.1A: Heating both the PCB and the resistor lead with a soldering iron

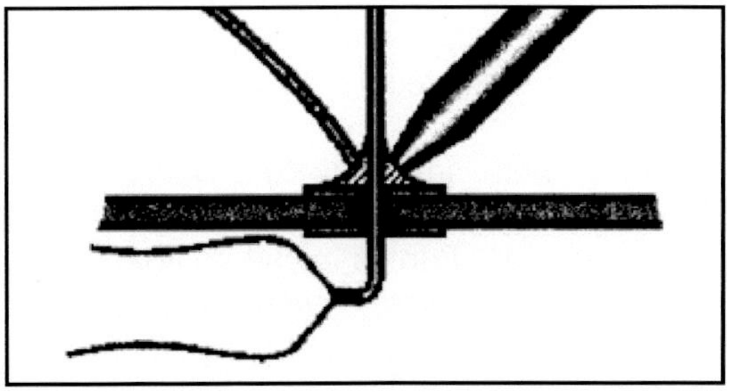

Fig 17.1B: adding solder to the heated surfaces

Begin with a mechanical joint

Solder is a soft material and it is best not to have a gap between the two surfaces to the joined. Try to enable the two surfaces to be in firm contact before the solder joint is made. A good mechanical connection will help to make a good solder connection.

Heat the joint not the solder

The job of the soldering iron is to heat up the surfaces to be joined rather than melt the solder. Try, whenever possible, to allow the surfaces to be joined melt the solder.

Begin the heating process by pressing the bit of the soldering iron firmly against the surfaces to be joined. On a printed circuit board jamming the point of the bit between the lead of the component and the pad on the board will heat both surfaces (**Fig 17.1A**). Two- or three-seconds contact will heat up the joint. A temperature-controlled soldering iron ensures that the joint is not over heated. The heated surfaces rather than the tip of the soldering iron should melt the solder (**Fig 17.1B**). This ensures that the joint is hot enough to allow a smooth flow of solder. Avoid melting the solder with the soldering iron tip and do not use the soldering iron to carry molten solder to the joint.

Inspect and tidy the joint

A bad joint, often called a 'dry joint', can usually be spotted by inspection. A good joint will be shiny and smooth. A dry joint will look dry and brittle. Trim off any excess wire using side cutters or diagonal cutters. Apart from being untidy, surplus bits of wire can cause short circuits, or a bridge between two tracks on a printed circuit board.

Successful soldering is really very easy to master. After a little practice it becomes second nature. But I remember what a kit manufacturer once told me, that 99% of the failures to successfully build his kits were caused by poor soldering.

Sorting out components

The amateur radio constructor will soon gather together a lot of very small parts. An amazing

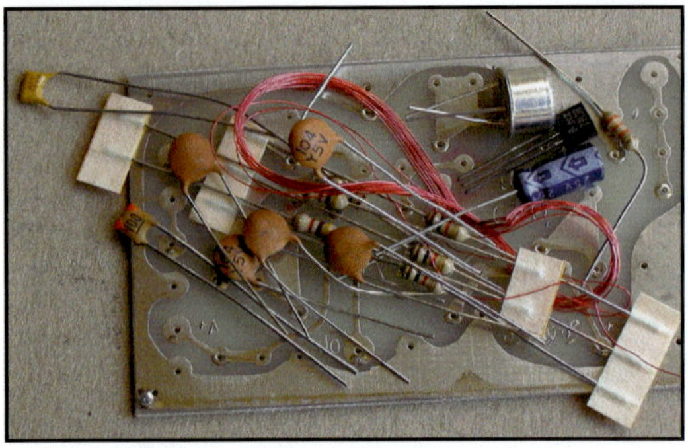

A set of parts for the OXO Transmitter kit

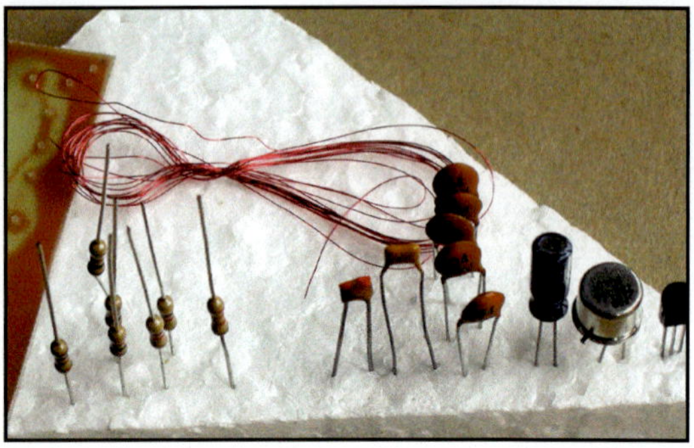

OXO Transmitter parts sorted according to value and secured on a piece of expanded polystyrene

amount of time can be taken up in simply sorting out the correct component for a particular place in the project. Having all the resistors in one box marked "resistors" is asking for hard work. Most wise constructors store their parts separated by their numeric value.

My stock of resistors and capacitors are all housed in small, cheap, manila envelopes. Each envelope contains components of the same value and the envelopes are housed in shoe boxes. I know that others like the more aesthetically pleasing small plastic drawers, but it takes an awful lot of drawers to separate every value of small component.

It is helpful to begin every project by gathering together all the required parts. It can be very frustrating to find out, part way through a session of construction, that a required component is missing. If the project is a kit, all the parts should be ready supplied for use.

The photograph shows the parts required to build an OXO Transmitter kit from Kanga Products. Before beginning work on the kit, it is a good idea to sort out all the parts into their numeric order. Some constructors like to place them in stored piles on the bench; An idea to be avoided as it is too easy to knock them off the table and have to scramble about on the floor locating tiny components. I find the easiest way is to stick the parts into a piece of polystyrene.

A piece from a polystyrene tile is convenient for this purpose, although I tend to save flat pieces of polystyrene packaging for the job. The leads of the components are pushed into the polystyrene to keep them secure in the working area. Notice that the types of component are separated and that the individual components are sorted by value. Ones of the same value are arranged in a row. This method of pre-sorting saves an immense amount of time and frustration when the component placement takes place.

The OXO is a simple, low component count, project so this way of working will save even more time for more complex projects.

Construction techniques

To make the QRP experience complete, the following chapters will describe some simple QRP projects, which can either be built from scratch using the components or from a kit. If the reader opts to buy the kit, the construction method relies on the printed circuit board supplied by the kit manufacturer. The more adventurous reader who decides to build from scratch will have to decide how to build the project. The obvious answer may appear to be a printed circuit board. The problem here is that the making of the board will probably be more complicated than building the actual project.

For 'one-off', small, amateur radio projects, the printed circuit board is not the best method of construction. Printed circuit boards are the ideal way to duplicate a circuit several times, but there are several simpler and more convenient ways to build a one-off project. The requirement is to wire the individual components to follow the electrical path of the circuit diagram. A printed circuit board looks neat and tidy but appearance is secondary to functionality. The aim is to get the project to work!

Any of the methods described below could be used with the projects described later in this book. The constructor can gather the parts, look at the circuit and decide how to build a working project. If the finished project does not look lovely, it can be hidden in a box. But even when exposed for all to see, it will at least look interesting.

Ugly Construction

The term 'ugly construction' was first coined by Wes Hayward, W7ZOI, and his son Roger, KA7EXM, in a joint article called 'The Ugly Weekender' in QST, August 1981. The article described a QRP transmitter which was built using point-to-point wiring over a piece of copper clad board.

The method is easy. The circuit is built over a piece of blank printed circuit board material, which acts as a 'ground plane'. All the components that are connected to ground on the circuit are soldered to the copper board using short leads. It helps to bend a small L-shaped foot on the end of the leads to be soldered to the board. Most of the non-grounded components are soldered to, and supported by, the grounded components as shown in the picture.

Where there are no convenient grounded components able to support other parts of the circuit, or where there are parts neither end of which are connected to ground, 'stand-offs' may be used. It is possible to buy commercial circuit board insulated stand-offs, but I use resistors. I choose resistors with a value high enough not to affect the operation of the circuit; usually 500k-ohms or higher. Extra capacitors can be used to support power lines around the circuit. A value of about 0.1μF (100nF) works well in radio frequency circuits and helps to remove any stray signal from the power line.

Another term sometimes used for ugly construction is 'Dead-Bug'. This comes from the common technique of mounting the transistors and integrated circuit 'chips' with their leads, or pins, pointing upwards. The chips do, indeed, look like dead bugs. This method keeps the connections on the device away from the ground plane and makes easier access for soldering to the pins. Dead-Bug constructors must be careful in identifying the pins because most pin-out diagrams are viewed from the top of the device.

Ugly construction does require having a supply of blank (un-etched) printed circuit board material. Blank copper clad board can easily be bought from electronics suppliers but it can be expensive. A better way is to look for off-cuts of printed circuit board material at amateur radio sales and events. The printed circuit board industry discards lots of good

Fig 17.2: Ugly construction

A one stage transistor amplifier built 'ugly style'

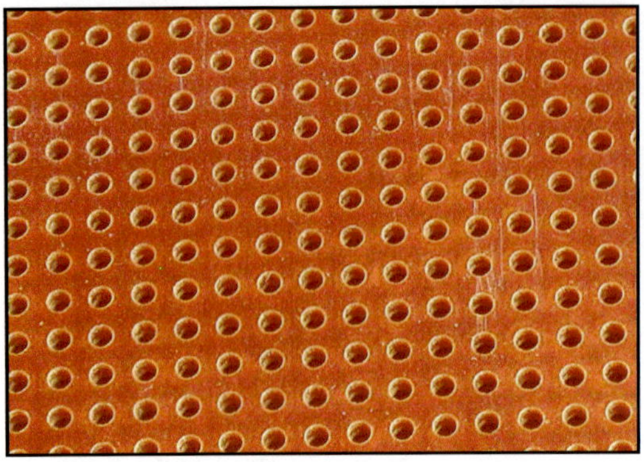

A piece of Perfboard

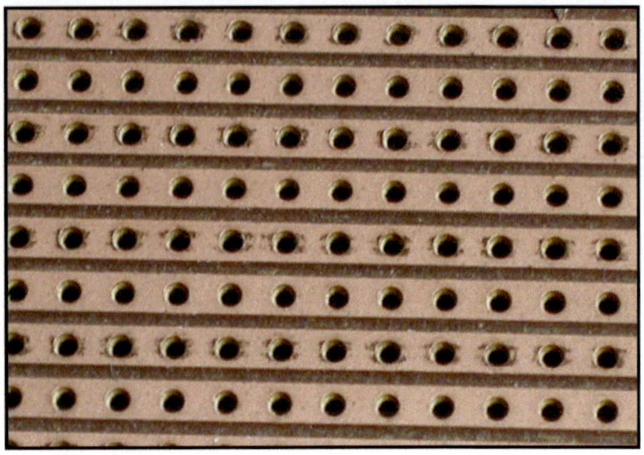

Veroboard which has copper strips should not be confused with Perfboard

pieces of blank board, usually the edges and ends trimmed from etched boards. It comes in all sorts of strange shapes and sizes but this surplus material is ideal for ugly construction.

Ugly construction technique does require some care and practice but over the years it has become my favourite way of prototyping circuits. It is a technique that can extend to complex projects. The famous 'W7EL Optimised Transceiver', featured in the ARRL Handbook, was built using ugly techniques throughout. If it had been built on a printed circuit board it would not have fitted into the case used for this back-packing radio.

Perfboard construction

Perfboard construction is a technique that closely mimics printed circuit board construction and many constructors like it because it has the same neat appearance. Printed circuit board layouts can be duplicated in Perfboard without the need to etch a copper clad board. Perfboard (also known as 'Matrix Board') is an insulated material that comes ready drilled with a matrix of holes spaced 0.1in apart. This is the standard pitch for integrated circuit pins and fits a whole variety of other electronic components. I recall once asking in a shop, that was part of a large stationary chain, for 0.1" spaced graph paper. The assistant assured me that no one used imperial spacing anymore and I would have to buy metric graph paper. I pointed out that everything that made the large electronic till, on the counter, work used 0.1in spacing.

Perfboard should not be confused with Veroboard. Both mediums look the same, except the under- side of Veroboard has a series of parallel copper strips joining row of adjacent holes. These are used to form tracks across the board. I shy away from using Veroboard for radio frequency work because the closely placed copper strips may cause problems at the higher frequencies. When building a project with Perfboard, the leads of the components are simply pushed through the holes in the board in the same manner as a conventional printed circuit board. It looks like a printed circuit board with extra holes.

An audio amplifier built on Perfboard

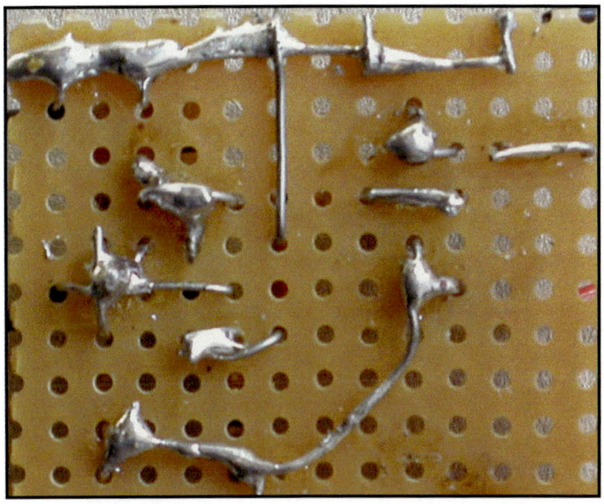

Typical connections under a Perfboard project

If the holes in the board do not correspond with the pins or leads on a component extra holes may be drilled. Some components may have pins too large to fit through the holes in which case they can be expanded using a twist drill as a hand tool. I have often

used a small conical grinding tool fitted to a 12-volt drill (Dremel-type) to modify Perfboard to suit component mountings. Perfboard is easy to saw to the required size, although it can be held in a vice and snapped into smaller pieces. The edges are then squared and smoothed using a hand file. To mount Perfboard, expand the size of corner holes to accept a small nut and bolt with a cylindrical standoff to support the board. The finished appearance can be very neat and follow a conventional printed circuit board style layout.

Wiring a Perfboard project is very easy. When the component leads are pushed through the board, the excess wire of the lead is used to make the component interconnections under the board. These leads act in the same way as the etched tracks on a conventional printed circuit board. If there is not enough excess lead length, or if the component has pins rather than leads, inter- connections can be made using tinned copper wire. It is also possible to use PVC sleeved wires when long tracks are required and there may be a risk of connecting wires touching. An insulated lead can also pass over another connecting wire should that be required to follow the circuit.

A minor problem in Perfboard construction is having to think 'upside-down'. The technique is a constant repetition of pushing in the parts on the component side and then turning over the board to make the interconnections. This means mirror reversing the layout each time the board is turned. The constructor needs care to keep the layout correct.

I usually begin Perfboard construction by inserting the main parts and then running wires beneath the board for the power and ground connections. Complex projects are possible using Perfboard. The more experience constructors sometimes mount components on both side of the board to gain extra space.

Pad construction

Pad construction is a method of mounting components on pads above a copper clad board. The idea is an old one. Nearly 20 years ago I wrote an article on a QRP transmitter called 'The SCD' which used this method of construction. In recent years it has enjoyed a revival in the USA under the name 'Manhattan Construction'. Chuck Adams, K7QO, awarding a prize to Jim Kortege, K8IQY, at a QRP construction competition at the Dayton Hamvention, first coined the phrase. He suggested the parts towered over the board like a Manhattan skyline.

It is a simple concept. Like ugly construction, the circuit is built over a copper clad board ground plane. In this method the components are soldered to small pads, made from printed circuit board material, which are attached on the ground plane. Usually the pads are attached to the ground plane using cyanoacrylate adhesive (Super Glue). Some pad constructors prefer to use doubled sided print- ed circuit board material for the pads which they attached to the ground plane by soldering the edges of the underside copper.

There are several ways to make suitable pads. One method is to use a metal nibbling tool to cut small pieces of copper clad material. A more usual method is to cut them from the board. Thin strips of copper clad material are easily cut using a sharp hacksaw and a mitre

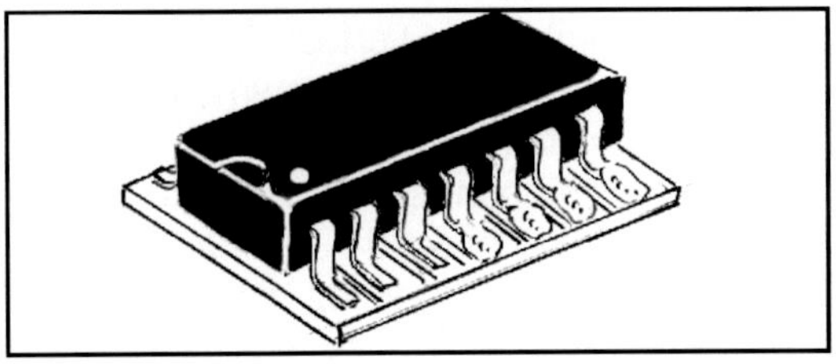

A pad IC carrier

A mounted pad

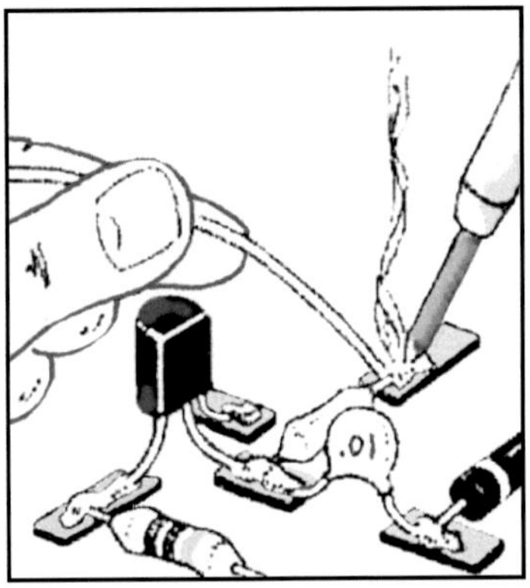

Soldering parts to the pads.

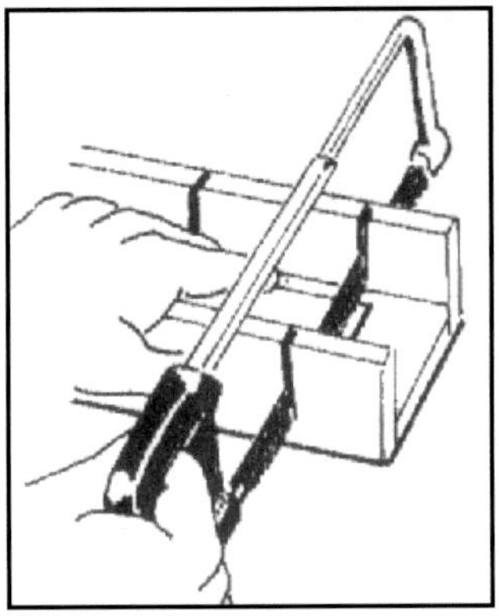

Cutting a copper strip

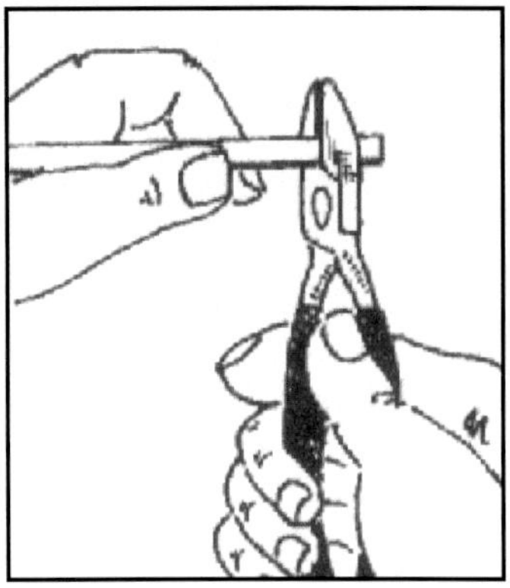

Snipping off pads

Drawings - Paul Harden, NA5N

block. The board is firmly held in the mitre block and careful sawing, parallel with the edge of the board, will produce a narrow strip.

The individual pads are cut from the strip using wire cutters or tin snips. Pads about 5mm square are ideal for a neat layout. These are a little small to cut so many constructors use larger pads. As each pad takes time to place and stick on the board, a vital part of pad construction is to lay out the board in advance. Gather all the required parts and lay them on the board to achieve the best circuit layout. The positions for the pads can be marked with a pencil.

When the layout is determined, the pads can be fixed, a few at a time, to the ground plane. A single drop of Super Glue is applied to the board and a pad is placed over the glue with tweezers and held down with a sharp point or cocktail-stick until it sticks. The constructor must observe all the usual precautions associated with cyanoacrylate adhesive. Although Super Glue does a good job in holding pads in place, they can be dislodged if too much heat is applied when soldering the parts in place. If pads need to be removed, usually they can be twisted off the board using needle nose pliers.

Integrated circuits can be used in pad construction by making a 'carrier' for the chip. This is a small piece of printed circuit board material cut to suit the footprint of the chip. A sharp hack- saw is used to cut breaks in the copper. One cut is made along the centre to divide the pins on other side of the chip and a series of parallel cuts form pads for the individual pins. The pins of the chip are bent at 90% from the body to sit firmly on the pads.

Some constructors simply prefer to use their integrated circuits mounted dead-bug style. In this method the top of the IC is glued to the board with the pins facing upwards. The connections are directly soldered to the pins after the style of ugly construction.

During the making of the board, the copper surfaces may become spoiled by handling, so before any soldering can take place they require cleaning. A toothbrush with a little alcohol-based solvent will do the job, although I prefer to use a light wire suede brush.

The components may now be soldered to the pads. Follow all the practices for making a good solder joint without over-heating the pads. It helps to bend the component leads so that a flat edge is presented to the pad. Remember, when placing component leads, to leave enough space for other leads, which may have to be added to the same pad.

Alternative chip carriers

An alternative method for integrated circuit (IC) chips is make a carrier from a small piece of veroboard and an IC holder. The first stage is to prepare a small piece of veroboard to hold the IC holder. For the 8-pin DIL (dual in line) chip in this example, the board requires four tracks and a width of eight holes. The tracks will act as solder terminals. This requires the tracks to be broken and this is done by using a vero cutting tool or a suitable drill bit to cut away the copper around each of four holes in a row. This should be the fourth row of holes from one end. When the IC holder is soldered in place, a portion of copper track with two holes will protrude from each side.

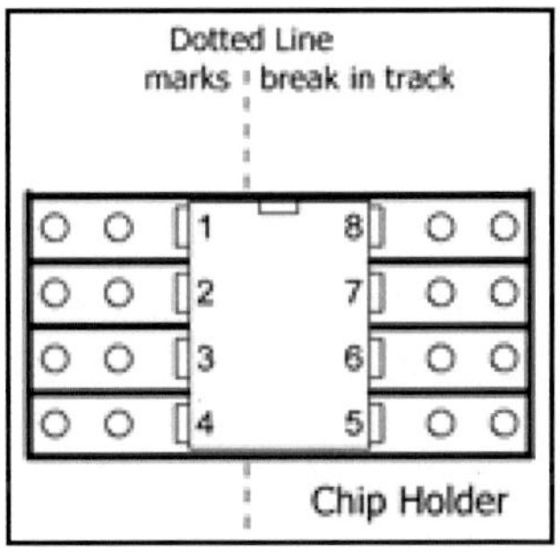

An alternative chip carrier

Commercial pads

Commercially made pads for Manhattan style construction are produced by a US company called QRPme (www.qrpme.com) owned by Rex W1REX. They are available in the UK from the GQRP Club. The pads are sold as 'MeSQUARES' in panels which have 300 square construction pads that are easily snapped apart and then glued down. Each pad is pre-tinned and has a solder mask and silk screen outline. QRPme also sell integrated circuit carriers in conventional and SMD footprints.

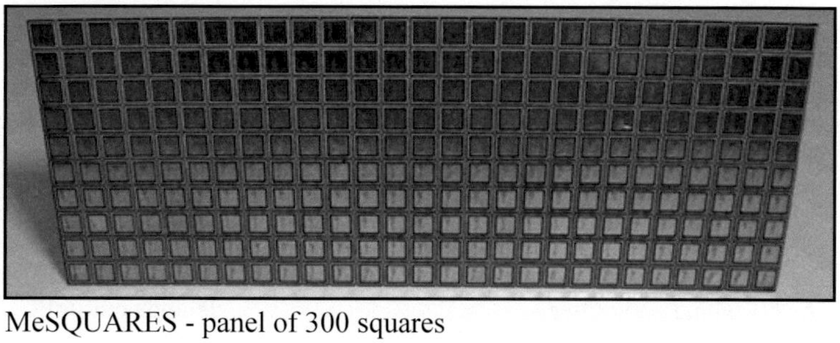
MeSQUARES - panel of 300 squares

A tip for mounting pads

The problem of gluing a pad in the correct place requires nothing more than a wooden cocktail stick (or a match stick) and a small piece of 'Blu Tack' stationary putty. Cut one pointed end off of the cocktail stick. Take a tiny ball of Blutack; about 5mm diameter and push it on to the blunt end of the cocktail stick. This makes a little tool for picking up a pad by pressing the Blu Tack onto the top (copper) side of the pad.

Turn the stick so that the bottom of the pad faces upwards and apply some super glue. It really only needs a very small amount of glue to secure the pad. The pad can be accurately placed using the stick and secured by pressing down on the stick for about 20 seconds. This is a quick and easy way to place the pad in the desired place and secure it to the ground-plane.

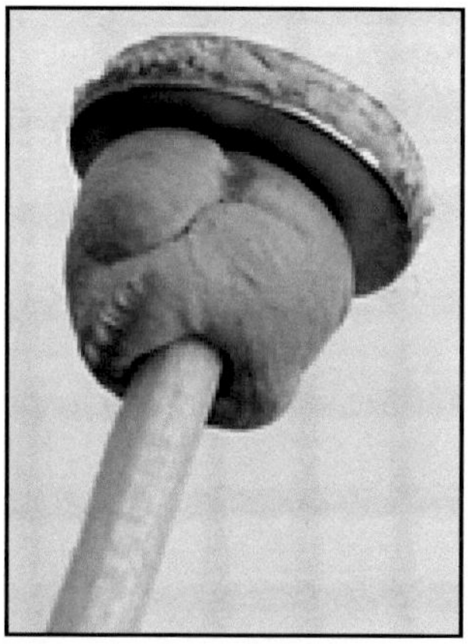
Mounting a pad with a wooden cocktail stick and 'Blu Tack'

Island construction

Island construction is a variant of pad construction. In this method small islands cut into the ground plane material replace the pads. The basic requirement is a tool to cut away the copper around the 'island' In the past constructors have used abrasive cutters attached to small 12V drills and home- made cutting tools.

Tony Fishpool, G4WIF, uses modified wood cutting drills to make his islands. The original drill (shown in the picture) is ground so that the centre point is the same height as the two outer cutting edges and one of the cutting edges is removed. The modified drill makes a small centre hole then cuts out a circle of copper to form a round island of copper on the board. This island is used in exactly the same way as the pads in pad construction. Care has to be taken when soldering to prevent solder flowing into the insulating circle around the island.

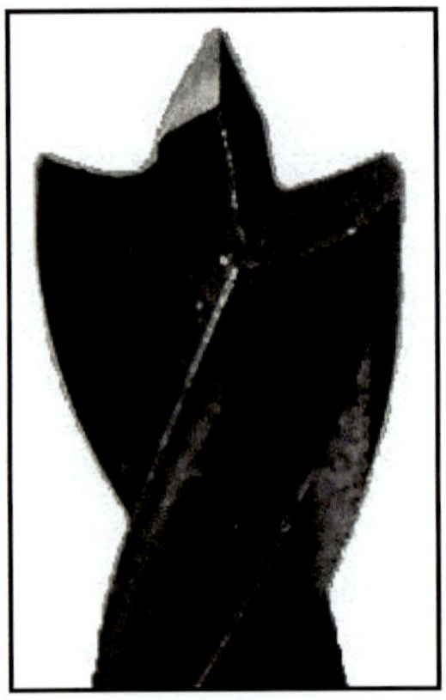

The original drill which was later modified to make an island cutting tool

The New Jersey QRP club has introduced an Island Pad Cutter. This tool is specially designed to cut neat circular islands in copper clad PCB material. George Heron, N2APB, of the NJ QRP Club says: "The islander pad cutting tool was envisioned by club member Dov Rabinowitz, ADOV, while sitting in a dentist's chair during a check-up. He later hunted down a manufacturer of little 5mm(0.2in)-diameter, diamond tipped, end mills. The NJ QRP then arranged a bulk purchase for the benefit of QRP homebrewers".

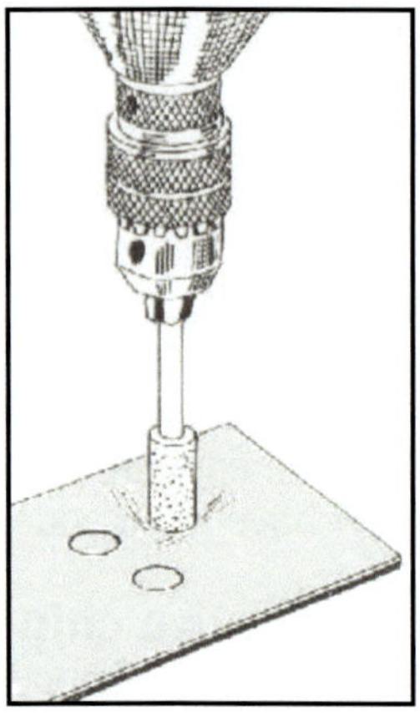

NJ Islander in action

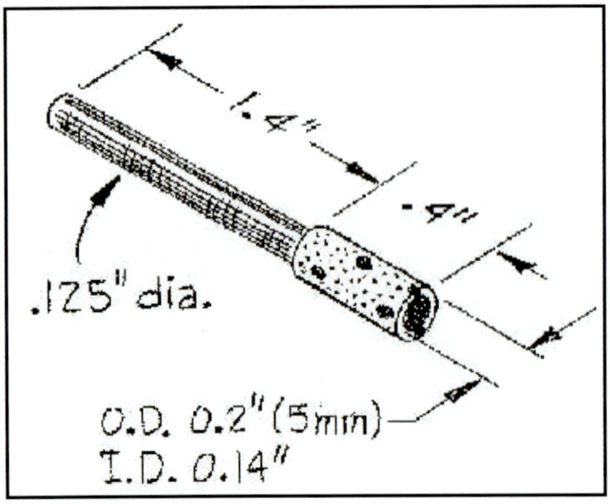

The NJ island pad cutter

The NJ islander mill can be used in a bench-mounted drill press or with a hobbyist's Dremel tool (12V drill) and tool stand. Like all printed circuit board drilling applications, a high drilling speed is required to prevent tearing. It is important to drill the islands perpendicular to the board to prevent 'bit walking'. A little soapy water on the cutting surface can help. Care is needed not to drill right through the ground plane material. Drill debris can be brushed away with an old toothbrush. The circle should be inspected for a clean-cut gap all the way around the island.

The islands provide a very neat and convenient series of pads on the copper clad board. Careful soldering is required to prevent surplus solder filling the cuts. The component leads are best mounted with a small 'L' leg to raise them clear of the insulated channel around the island. Because the islands are flush with the board, surface mount parts may also be used.

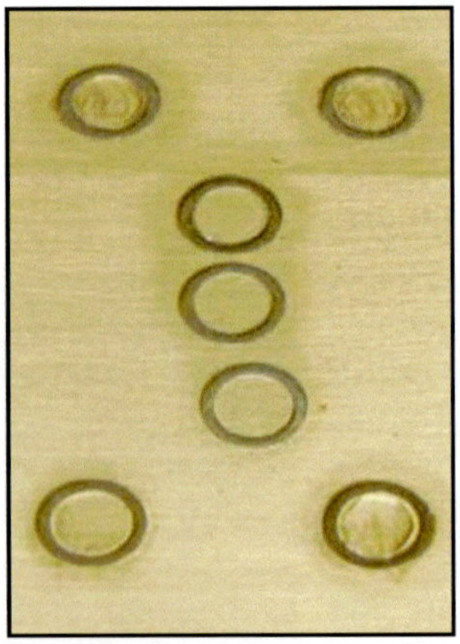

Islands cut in PCB material

Integrated circuits can be added to the board by using glue down IC carriers as described above, or the constructor could use the dead-bug tech-

Wire wrap IC holders mounted over cut islands

Example circuit with copper tape supplementation underneath to make a better earth rail

The New Jersey QRP Club produced a little project to introduce the cutter, the K8IQY Islander audio amplifier

nique. An article by ADOV in the NJ QRP Club journal, QRP Homebrewer suggests using wire-wrap sockets, with their long leads splayed out to reach rows of islands. The Islander Pad Cutter is available in the UK via the GQRP Club.

Copper tape methods

An interesting variation on the do-it-yourself PCB approaches is to use copper tape, which can be cut and shaped, and then stuck down to an insulating substrate. The copper tape used is that commonly applied in plumbing, or for decorative purposes, and has an adhesive on one side. This can be used to supplement Perfboard construction, or to turn a Veroboard-based circuit into a double-sided PCB, in effect, or to supplement the existing copper tracks.

The copper tape is usually adhesive enough to stay in position, with the components on the top side (as shown) helping to anchor it in place permanently. It also has the interesting property that it can be stuck to vertical- as well as horizontal surfaces, or even curved surfaces. This allows greater circuit density within a small box, for example.

Putting it in a box

The finishing touch to any project is providing a case and the necessary hardware; controls, switch- es and knobs. These can be expensive. A beginner to radio construction may be surprised to find that the box and hardware can cost more than the entire electronic content. The frugal constructor will collect knobs, switches, meters and other items of hardware whenever they appear cheaply, or cull them from discarded equipment.

Earlier I mentioned the quirky practice of QRP constructors to use small tins, Altoids mint tins being a cult favourite. I have often bought surplus items at junk sales if the case looks useful, and discarded most of the contents. Some constructors make simple cases from printed circuit board material. It is easy to work and can be covered or painted.

For those who prefer a new case and do not have equipment or skills in sheet metal working, some inexpensive off-the-shelf cases are available. The simplest, and often cheapest are the 'clam-shell' type boxes. These have an integrated base, front and back panel, with a top and side panel which fits over the top. They are usually made from sheet aluminium and are easy to work with hand tools.

Working aluminium is easy because it is a relatively soft metal. But that property itself can be the cause of problems. Over the years I have

A simple 'clam-shell' type box

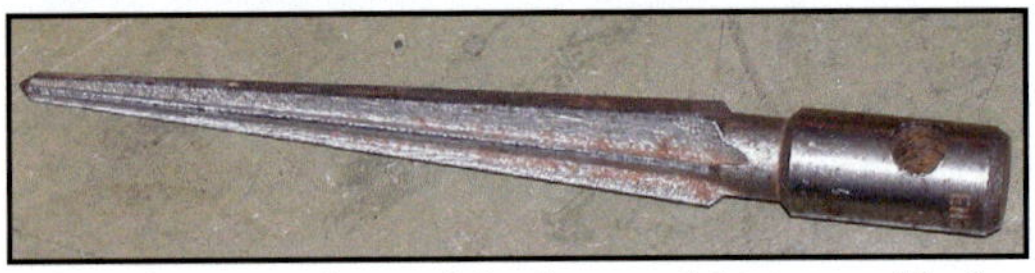

A hand reamer is useful when making round holes in aluminium

How the two parts of the box fit together

found that drilling a round hole is aluminium is very difficult. Unless the panel is held in a rigid support, the drill bit will wander and produce a distorted hole. The answer is to use a hand reamer. Holes are drilled smaller than the required size and reamed up until the size is correct. This has the additional advantage of being able to use one drill size for several sizes of hole. The completed box can be painted and labelled. Experience has taught me that paint- in bare aluminium is difficult so I use false front panels.

Making simple scratch-proof front panels

When the front panel has been drilled and checked, remove all the controls. Place it face down on to a sheet of thin card or good paper (it can be coloured paper or card). Using a ballpoint pen, mark around the inside of all the holes; this will be the reverse of the front panel. Press firmly so the marks show through to the front. If the piece of paper/card is larger than the panel, mark around the out- side also. Turn over the paper/card to see the front with the indented marks showing. Add the required legends. This can be done with a pen, stencils, rub-down lettering, computer drawing soft- ware . . . in fact anything that can write on to paper or card.

Glue the paper/card front to the front panel, holding it up to the light to get the positions of the holes correct. I use a glue stick. Cover the front with 'library film'; the clear sticky-backed plastic often used to cover books. This can be tricky as it is easy to produce wrinkles. I cut the sticky-backed plastic to over size and introduce it to the panel from one end, smoothing it out as I proceed. Trim with a little excess all round. The excess is smoothed around the edges of the panel to give a more secure fit.

Using a sharp modelling knife with a point, cut out the holes. Press the knife against the inside of the hole and gently work it round. Add the controls, meters etc with care. Screwing up control nuts too tightly can wrinkle the plastic film. Adding a washer on the outside is a good idea, if the knob will hide it. This gives a clean scratchproof, wipe- clean, surface. It can also hide nasty marks on the panel and even extra holes from a former use of the box.

Home-made front panel made from card - as described above

Chapter 18

Toroids, Filters and Inductors

A casual glance at many QRP projects will spot inductors (coils) wound on small circular, Polo-Mint like, cores. These are toroidal cores much favoured in QRP radio design, often simply called 'toroids'. Some inexperienced constructors are daunted by the thought of winding their own inductors but it is really very easy. Having said that, as a sometime writer of practical amateur radio articles, I keep my fingers crossed whenever I include home-wound inductors in a design. Constructors can get them wrong and then write to me complaining that the project does not work. What follows is a short primer in the use of toroidal cores for home-wound inductors.

Usually QRP radio designs use the range of cores manufactured by Micrometals, often sold by Amidon Associates of California. They have the advantage of being compact and easier to add windings to than the cylindrical type of coil former. They tend to require fewer turns to achieve a specific inductance. The main advantage is that toroidal inductors are self-shielding; most of the magnetic flux lines are contained within the shape. The flux density is almost uniform over the whole winding and they do not require mounting inside a screening can.

The commonly used toroidal cores come in two types; those made with ferrite material and those made using iron powder material. The ferrite types have a higher permeability. The usual first question is why not use ferrite material for all the cores; the greater permeability would mean having to wind less turns to achieve a required inductance? The simple answer is, the higher the permeability of the core, the less constant the inductance will be at higher frequencies. So, ferrite cores are usual for broadband transformers and radio frequency chokes and iron powdered cores for tuned circuits.

Toroidal cores are designated according to their material (iron powder or ferrite), their diameter and the variety of material. A core designated as T-50-2 is a powdered iron core

A low-pass filter using toroid inductors

A selection of toroid cores

(T = powdered iron) with an outer diameter of 0.5 inch (50 = 0.5in) and uses a '2-mix' of powdered iron. A core designated as FT-37-43 is a ferrite core (FT = ferrite) of 0.37in outer diameter using '43 mix' ferrite.

This may sound complex and indeed the full technology of toroidal cores is detailed but thankfully only a few types of core are commonly used in QRP circuit design. QRP constructors are only likely to meet the following cores.

Commonly used cores

Powdered iron

The two common powdered iron types are 2-mix (painted red) and 6-mix (painted yellow). A simple rule of thumb is to use 2-mix in tuned circuit up to 10MHz and 6-mix in tuned circuits over 10MHz. The core usually encountered are:

T-68-2, T-68-6: Powered iron core – diameter 0.68in – 2-mix (red) or 6-mix (yellow)

T-50-2, T-50-6: Powered iron core – diameter 0.5in – 2-mix (red) or 6-mix (yellow)

T-37-2, T-37-6: Powered iron core – diameter 0.37in – 2-mix (red) or 6-mix (yellow)

Ferrite

According to application, the two most common ferrite types are 43 material and 61 material. The cores usually encountered are:

FT-50-43, FT-50-61: Ferrite core – diameter 0.5in – 43 material or 61 material

FT-37-43, FT-37-61: Ferrite core – diameter 0.37in – 43 material or 61 material

Other types do appear in amateur radio design but the above are, by far, the most common. Also note that sometimes a hyphen is dropped in the designation and, for example, FT-37-2 might be writ- ten as FT37-2.

Winding toroidal inductors

Instructions for winding an inductor will usually give the type of core, the number of turns required and probably the size (gauge) of wire required. The wire is enamel coated to prevent turns short- circuiting. If the instructions are above averagely helpful, they may include the length of wire required to make the coil. It is easier to work with a length of wire not attached to a bobbin so having the required wire length ready is helpful.

An inductor wound on a toroidal core

Experienced coil winders probably know how much wire they will need for a specified number of turns on a core; there are even some charts which give suggested lengths. The simple way is to put a few turns on the core, remove them and measure the length used and then do a multiplication to work out the length for the required number of turns.

The photograph below shows two windings on a core. The winding is made by passing the wire through the centre of the core. Allowing enough free wire to make a connection to the circuit board, or other component, begin making

Two turns on a toroidal core (two passes through the centre)

A completed inductor.

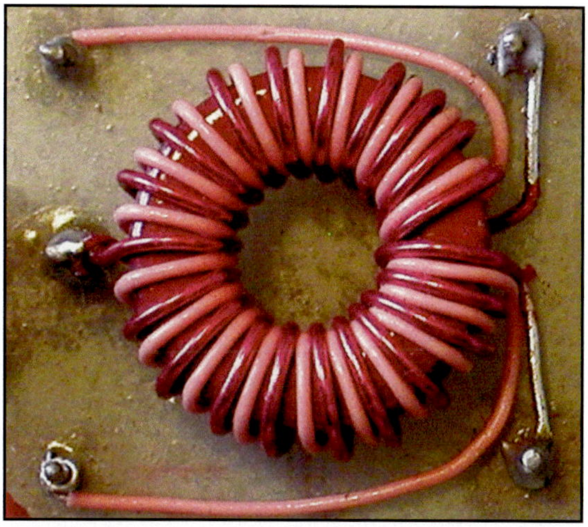

An inductor with an interleaved second winding and a tap

turns through the core. The important rule is: every time the wire passes through the hole equals one turn. Count as the turns are made. If the number of turns is forgotten, usually when the phone rings half way through the process, the wires passing through the centre can be counted.

It is normal to wind the coil so that the winding occupies about 80% of the core. Attempt to dis- tribute the turns evenly around the core. Squeezing or opening up the winding can alter the inductance. If the beginning and ending turns are too close together this will add capacitance to the circuit.

The photograph below shows a completed inductor on a toroidal core. One of the leads has been prepared for mounting. Before the inductor can be soldered in place, the enamelled coating on the wire must be removed and the bare copper tinned with solder. Some types of enamelled wire are designed so that the enamel melts away when a hot soldering iron is applied. Others will require the enamel to be removed by scraping it away with a modelling knife or emery paper. Heating the bared copper and allowing a layer of solder to flow over the surface will tin the end of the wire.

Some inductors may require other windings to be added to the same core to form a transformer for example). These can be wound over the main winding. In some cases, the added winding can be interleaved with the original winding. The photograph shows a second winding interleaved with the main winding. The main winding is centre-tapped. To make a tap on a winding, pull out a small loop of wire at the appropriate place on the winding.

Remove the enamel from the loop and twist the loop into a tight 'pigtail'. Tin the twisted loop with solder and a tap is formed.

Bifilar and trifilar windings

Two relatively common forms of inductor winding, called 'bifilar' and 'trifilar' windings sometimes cause the constructor distress. These are windings in which two wires (bifilar) or three wires (trifilar) are twisted together and wound as one wire. Radio frequency transformers often use this form of winding to enable good coupling between two circuits. They are a little more tricky than conventional single wire windings but easy enough if care is taken. The usual problem arises when deciding to join which wire to where.

The best way to understand these windings is to look at a simple example. The circuit diagram shows 4:1 balun; a balanced to unbalanced transformer. An arrangement like this may be used to match a 50-ohm antenna output to a balanced feeder line.

Fig 18.1 shows two windings (1 to 2 and

3 to 4). These will be wound using two wires twisted together.

End 1 and end 3 are marked with dots; the usual way to show the beginning of each winding. When wound, 1 and 3 will be at one end of the core with 2 and 4 at the other end of the core.

Prepare two lengths of wire - each should be about one and half times longer than is required to make the winding. I usually stretch the two wires side by side and tie a knot in each end to secure them together. I fasten one end in a vice, or the jaws of a Third Hand and thread a pencil through the loop of the tied wires at the free end. By rotating the pencil, it is possible to twist the two wires together. Attempt to get about 4 or 5 twists per centimetre. The twists may have to be stroked down by hand to get an even distribution. The twisted wires are now used as one wire.

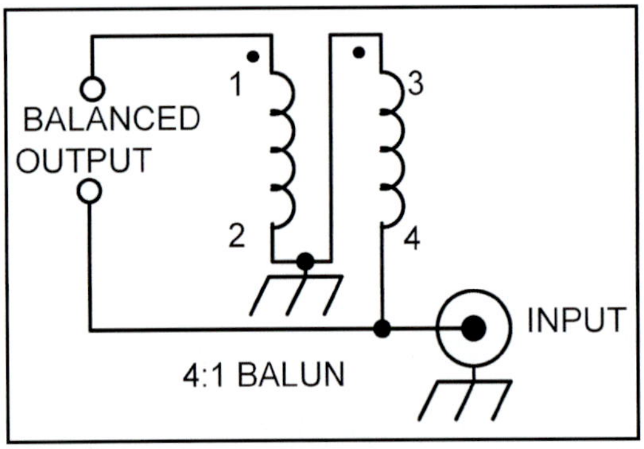

Fig 18.1: Circuit of a typical balun wound on a toroid

Wind the coil using the twisted pair of wires, each insertion through the hole counting as one turn. It will become one winding with four ends. Remove the enamel from all four ends and tin the bared cop- per. Now comes the bit that some people get wrong; sorting out which wire goes to where in the circuit. The photograph shows the completed bifilar winding. In this example I have used two wires with different coloured enamel coatings to show the two windings more clearly. If you can get hold of wires with different coloured coatings, it is not a bad idea to use them. It certainly helps to identify the connections. At one end of the core we have the beginning of both windings (1 and 3) and at the other end of the core we have the end of both windings (2 and 4).

Following the circuit diagram in Fig 18.1, wire 4 will be joined to the 50-ohm input point. Wire 3 (the beginning of winding 3-4) will be joined to wire 2 (the end of winding 1-2 and the joined wires connected to ground. The remaining wire, 1, is connected to the balanced output.

How does the constructor work out which wire end is which, if both wires are the same colour? This requires a continuity tester; a device to show if there is a direct electrical circuit be-

A bifilar winding with the ends numbered

Another bifilar wound transformer

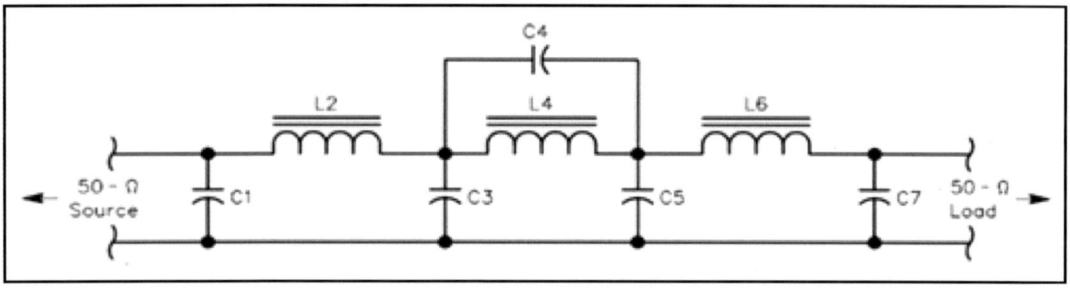

Fig 18.2: Circuit diagram of the W3NQN CWAZ low pass filter

TABLE 18.1: CWAZ 50Ù LOW-PASS FILTERS DESIGNED FOR 2ND-HARMONIC ATTENUATION

BAND m	START FREQ MHz	C1/C7 pF	C3/C5 pF	C4 pF	L2/L6 µH	L4 µH	F4 MHz
-	1.00	2986	4556	680.1	9.377	8.516	2.091
	1.80	1659	2531	378	5.21	4.73	3.76
160		1450+220	2100+470	330+47			
		1500+150	2200+330				3.78
	3.50	853	1302	194	2.68	2.43	7.32
80			1150+150				
		470+390	1200+100	150+47			7.27
40	7.00	427	651	97.2	1.34	1.22	14.6
		330+100	330+330	100			14.4
30	10.1	296	451	67.2	0.928	0.843	21.1
		150+150	470	68			21.0
20	14.0	213	325	48.6	0.670	0.608	29.3
		220	330	47			29.8
17	18.068	165	252	37.6	0.519	0.471	37.8
		82+82	100+150	39			37.1
15	21.0	142	217	32.4	0.447	0.406	43.9
		150	220	33			43.5
12	24.89	120	183	27.3	0.377	0.342	52.0
		120	180	27			52.4
10	28.0	107	163	24.3	0.335	0.304	58.5
		100	82+82	27			55.6

The upper capacitor values in each row show the calculated design values obtained by dividing the 1MHz capacitor values by the amateur-band start frequency in megahertz. The lower standard-capacitor values are suggested as a convenient way to realise the design values. The middle capacitor values in the 160- and 80-metre-band designs are suggested values when the high-value capacitors (greater than 1000 pF) are on the low side of their tolerance range. The design F4 frequency (see upper value in the F4 column) is calculated by multiplying the 1MHz F4 value by the start frequency of the band. The lower number in the F4 column is the F4 frequency based on the suggested lower capacitor value and the listed L4 value.

tween two points. A battery and bulb would do it, but every good constructor has a multimeter. The multimeter will have a range, or several ranges, for measuring resistance. Select the lowest 'ohm range' and test the wires at either end of the coil with the meter probe to identify the beginning and end of each wire.

The second picture shows a bifilar wound balun transformer with the leads prepared ready for use. It also shows the beginning of one winding joined to the end of another winding (3 and 2 in our example above). This transformer is wound with two wires having the same coloured enamel coating. Trifilar transformers are wound in exactly the same way, except that three wires are used rather than two. The beginnings and endings of the windings can, again, be identified with an ohmmeter. It is even more fun sorting out which wire goes where in the circuit!

Harmonic filters for QRP transmitter output

Although by their very nature, QRP transmitters radiate less power, the output from such a transmitter does require adequate filtering. Usually to keep the circuit compact, these transmitters have a final stage running in Class C and being driven hard with RF power. Of itself, this is a recipe for high harmonic output and a well-designed low pass filter is essential.

For many years QRP transmitter constructors have used low pass filters based upon the excellent work done by Ed Wetherhold, W3NQN (SK), in the 1980s. His set of filters based on standard capacitor values have been superseded by later work to improve the second harmonic attenuation. In an article in the February 1999 edition of QST magazine, W3NQN offered data on building second harmonic optimised low pass filters. The design, shown in **Fig 18.2**, follows his former seven-element circuit but adds an extra capacitor across the centre inductor.

Table 18.1 shows the values for the capacitors and inductors for the HF amateur bands. Note that in most cases the capacitance values are made up using more than one standard value capacitor. The capacitors are connected in parallel to achieve the desired capacitance.

The inductors are all wound on toroidal cores in the popular Micrometals range. The wire gauge is not critical. Simply use the gauge that will fit well on the core. The target is to wind an even coil on the core to occupy about three-quarters of the available space. If the opposite ends of the winding are too

TABLE 18.2: PRACTICAL EXAMPLES FOR <10W TX OUTPUT

Band	C1/C7 pF	C3/C5 pF	C4 pF	L2/L6 turns/core	L4 turns/core
160m	1500+150	2200+330	330+47	33t T50-2	31t T50-2
80m	470+390	1200+100	150+47	26t T37-2	25 t T37-2
40m	330+100	330+330	100	21t T37-6	20t T37-6
30m	150+150	470	68	18t T37-6	17t T37-6
20m	220	330	47	15t T37-6	14t T37-6
17m	82+82	100+150	39	13t T37-6	13t T37-6
15m	150	220	33	12t T37-6	12t T37-6
12m	120	180	27	11t T37-6	11t T37-6
10m	100	82+82	27	11t T37-6	10t T37-6

The inductors are all wound on toroidal cores in the popular Micrometals range. The wire gauge is not critical. Simply use the gauge that will fit well on the core. The target is to wind an even coil on the core to occupy about three-quarters of the available space. If the opposite ends of the winding are too close this will introduce extra capacitance.

TABLE 18.3: USING WOUND TOROIDS		
BAND MHz	TURNS ON TOROID	C pF
18	55t on T50-2	470
3.5	45t on T50-2	200
7.0	36t on T37-2	100
10.1	35t on T37-6	68
14.0	30t on T37-6	47
18.07	26t on T37-6	39
21.0	24t on T37-6	33
24.89	25t on T37-6	22
28.0	24t on T37-6	18

TABLE 18.4: USING AXIAL CHOKES		
BAND MHz	AXIAL FIXED INDUCTOR	C pF
1.8	47u.H	150
3.5	39uH	50
7.0	l0uH	50
10.1	5.6uH	47
14.0	2.7uH	47
18.07	2.2uH	33
21.0	1.8uH	33
24.89	1.8uH	22
28.0	1.2uH	27

close this will introduce extra capacitance.

Table 18.2 shows actual values for practical filters using inductors wound on toroidal cores. Wire gauge is not critical. Use a wire size to fill the core about three-quarters of full circumference. The number of turns has been rounded to the nearest whole number from the calculated value.

Tuned circuits to hit the bands

With very little fiddling these values produce tuned circuits on the amateur bands.

Table 18.3 shows values for winding inductors on toroids to reach the desired band. Adjustment is best achieved by making part of 'C' a trimmer capacitor in parallel with some fixed capacitance

Table 18.4 shows some values for using axial chokes. These tuned circuits use standard value fixed axial inductors so the design is a bit subjective! Values can vary, the Cs are a compromise and Q can be low, but ideal for small spaces. They will need adjustment by making part of 'C' a variable trimmer.

Ready designed receiver bandpass filters

This series of bandpass filters is designed for the input tuning of amateur band receivers based on data from GQRP Club member Stefen Niewiadomski from the GQRP Club website.

The filters are designed for a 50-ohm input and output impedances and may be cascaded for better roll-off outside the passband.

The coils also have a tap on the tuned winding which is useful for matching into the input circuit of the order of 1kΩ.

Formerly the designs used off-the-shelf Toko coils. These are no longer made but Spectrum Communications (http://www.spectrumcomms.co.uk/) now import their own range of coil most of which are compatible with the Toko range.

The inductors are variable [core tuned] but the mean inductance is given for matching to other inductors.

Some VFO values

The Colpitts VFO is commonly used in HF amateur radio circuits. A typical Colpitts oscillator circuit is shown in **Fig 18.3**.

L with VC/C4 form the tank (tuned) circuit coupled to the oscillator via C3. C2/C1 are the capacitive feedback tap.

Shown in the table below are some 'rule of thumb' values for this popular VFO.

The value for VC/C4 is both capacitors in

SPECTRUM COMMUNICATIONS COILS (WITH THE NEAREST TOKO EQUIVALENT)													
L µH	**TYPE**	**TOKO**	**CpF**	**F MHz**	**T1-2**	**T-3**	**T1-3**	**T4-6**	**Q**	**Z pri**	**Z(1-2)**	**Z(2-3)**	**Z sec**
0.6	0u6H			50	3	3	6	1	75	9424	2356	2356	262
1.2	1u2L			28	6	2	8	1	85	17944	10093	1122	280
1.2	1u2H	3335		28	4	4	8	2	85	17944	4486	4486	1121
1.7	1u7H	4612		14	0	0	11	3	45	6729	0	0	500
2.6	2u6DC	586	82	10.7	0	0	12	0	100	17479	0	0	0
2.6	2u6L			10.7	7	7	14	1	85	14857	3714	3714	75
2.6	2u6LC	1509	82	10.7	7	7	14	1	85	14857	3714	3714	75
2.6	2u6FC	3892	82	10.7	7	7	14	2	80	13983	3495	3495	285
2.6	2u6HC	3893	82	10.7	7	7	14	3	80	13983	3495	3495	642
5.3	5u3L			7	15	5	20	1	85	19814	1238	11145	49
5.3	5u3H	3334/7		7	10	10	20	4	85	19814	4953	4953	792
9	9u0H	2027		5	3	0	28	5	70	19792	227	0	631
11	11u0L			3.6	23	7	30	2	80	19905	1083	11699	88
23	23u0L	~2926		1.9	37	11	48	3	50	13728	720	8156	53
45	45u0L	3333		1.9	42	14	56	3	60	32232	2014	18130	92
90	90u0L			1	57	19	76	5	65	36756	20675	2297	159

Base connections: *Suffix L is low impedance secondary and suitable for aerial connection. Suffix H is high impedance secondary suitable for interstage use and Hartley oscillators. Suffix D is discriminator, Suffix F is 300 Ohm filter matching. Suffix C includes a capacitor in parallel with the main winding.*

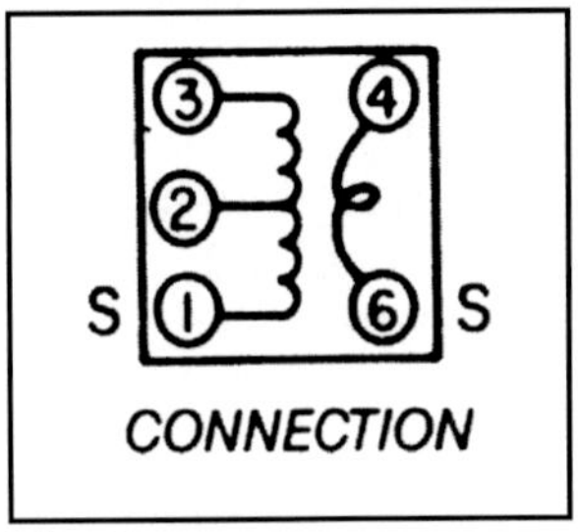

The Spectrum Communications Coil

BANDPASS FILTERS FOR THE AMATEUR BAND USING SPECTRUM 10MM COILS					
Band	**Tr1 / Tr2**	**Coil Marking**	**Core Colour**	**C1 / C3**	**C2**
1.8	45u0L	45u0L	Red	160p	12p
3.5	45u0L	45u0L	Red	39p	3p3
7.0	5u3L	5u3L	Yellow	100p	8p2
10.1	5u3L	5u3L	Yellow	47p	6p8
14.0	5u3L	5u3L	Yellow	27p	3p3
18.07	2u6L	KAXS1509	Blue	33p	3p3
21.0	2u6L	KAXS1509	Blue	22p	3p3
24.89	1u2L	FCPL3335	Pink	39p	3p3
28.0	1u2L	FCPL3335	Pink	27p	3p3

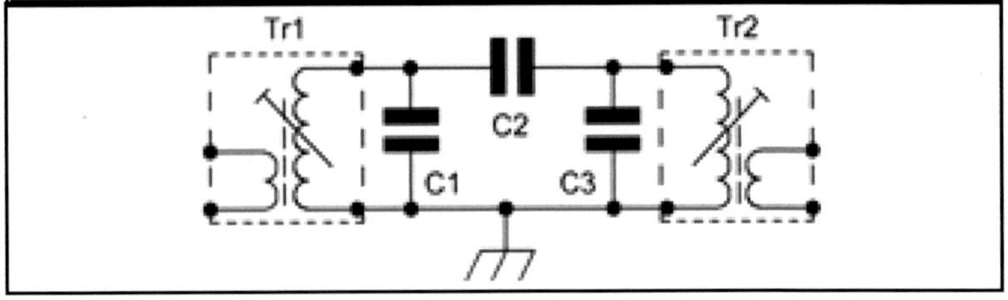

parallel. Adjust the relationship between VC and C4 to obtain the desired tuning range. If L is not adjustable, C4 may be made from a fixed value in series with a smaller trimmer capacitor to pull the circuit to the desired range.

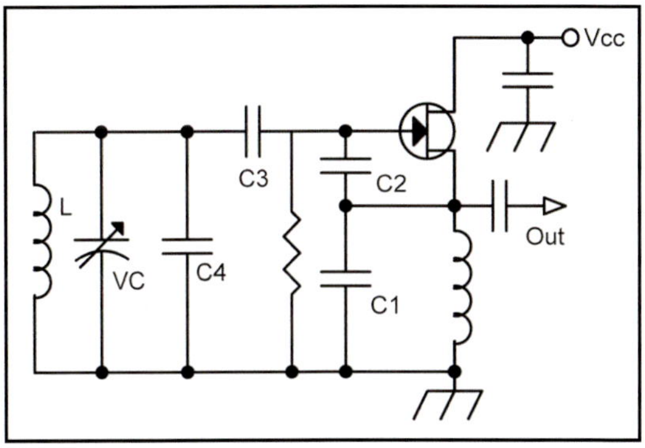

Fig 18.3: Typical circuit for the Colpitts oscillator

TYPICAL VALUES FOR L, VC/C4, C3, C2 AND C1

Band	L	VC/C4	C3	C2	Cl
1.8MHz	lluH	200	1000	2200	2200
3.5MHz	5.5uH	120	510	1000	1000
5MHz**	3.9uH	75	410	680	680
7MHz	2.9uH	50	250	510	510
14MHz	1.4uH	30	120	250	250

Note that C1/2/3/4 should becapableof good temperature stability. Polystyrene, COG or NPO capacitors are advised. Capacitor vatues are in pF.

*** The values may be scaled to suit other frequencies. Shown here is an example of a 5.0 to 5.5 MHz VFO for use, with the typica 19MHz IF superhet*

INDUCTANCE, FREQUENCY, AND CAPACITANCE GUIDE FOR SPECTRUM COILS

0.6uH	F (MHz)	70	50	29	21	
	C pF	8p2	18	56	82	
1.2uH	F MHz	29	24	21	18	
	C pF	27	39	47	68	
2.6uH	F MHz	21	18	14	10.7	9
	C pF	22	33	47	82	120
5.3uH	F MHz	14	10	7	3.8	
	C pF	27	47	100	330	
11uH	F MHz	10	7	3.65	1.84	
	C pF	22	47	180	680	
23uH	F MHz	7	5.25	3.65	1.82	
	C pF	22	39	82	330	
45uH	F MHz	5	3.8	1.94	1	
	C pF	22	39	150	560	
90uH	F MHz	3.6	1.85	1	0.53	
	C pF	22	82	270	1000	

Chapter 19

Practical Projects – Test Gear

Many QRPers derive immense satisfaction from using equipment they have built themselves. For some this can be a sophisticated QRP transceiver, but for many the appeal of QRP radio construction is the simplicity of what can be built and gainfully used. Without a dedicated workshop or great technical expertise, a lot of enjoyment may be had from making QRP equipment. The feedback I have received over many years, suggests that the favourite QRP projects are those which can be built in an evening or two, using inexpensive parts and with a high probability of working first time. Not everyone wants to become a technical radio expert but many enjoy the satisfaction, and therapy, of building modest QRP equipment.

The projects I describe in the following chapters are simple enough for the novice builder or any- one who can master soldering and recognise a few basic radio components. The technology is simple and the cost is low. Any of the projects can be built by gathering the parts and assembling the circuit using non-printed circuit board techniques but most of them are also available in kit form. These projects would make a good starting point for world-be constructors who may want to move on to more sophisticated circuits or more advanced kits. The invitation is to buy a few electronic components, gather a few basic hand tools, plug in the soldering iron and have a lot of fun.

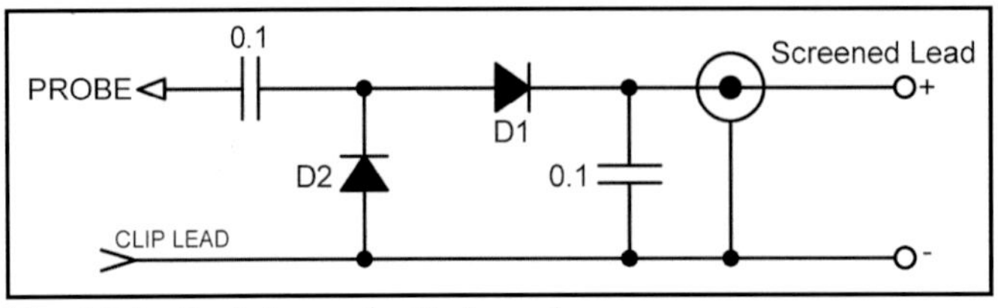

Fig 19.1: Peak-to peak diode probe

Before we get down to the more exciting things, which transmit or receive, it is important to have a few little items for testing future projects. They may not be very exciting in themselves but the constructor will be glad to have them at hand when later projects are being built.

Earlier in this book, the multi-meter (analogue or digital) was cited as an essential test item in the amateur radio workshop. This is usually capable of measuring DC (direct current) or AC (alternating current) voltage and current. The AC ranges on an average multi-meter are not capable of measuring RF (radio frequency) voltages. To do this, one of the simplest of all items of test equipment is required – the RF probe.

RF probes

The task of an RF probe is to convert a radio frequency (RF) voltage into a DC voltage so that it can be measured on a multi-meter. I have described the RF probe as a "crystal set on a stick" which comes in two main flavours: peak-to-peak probes and RMS (root mean square) probes. In practice, what the constructor requires, usually just checking the level of a radio frequency signal, can be done with a peak-to-peak RF probe and a DC voltmeter.

Fig 19.1 shows the circuit of a peal-to peak RF probe. It uses a couple of diodes to convert the radio frequency signal to a DC voltage. Crystal set builders will recognise what is being done. The input is picked off the tested circuit with a metal probe - hence "crystal set on a stick".

The two diodes are handling radio frequencies and ought to have a low forward voltage drop.

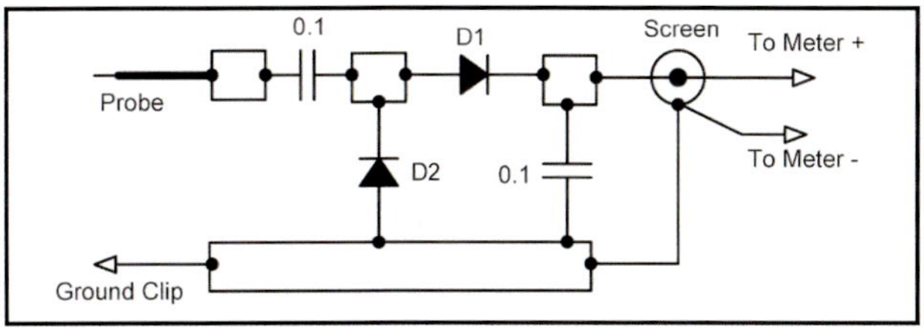

Fig 19.2: RF probe layout

A 'copper island' RF probe

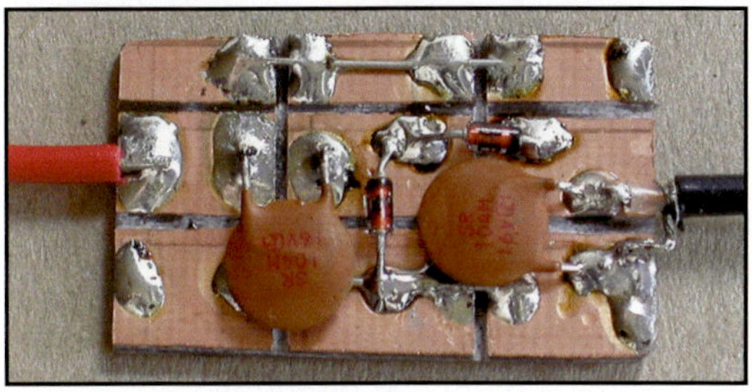

A 'matrix board' RF probe

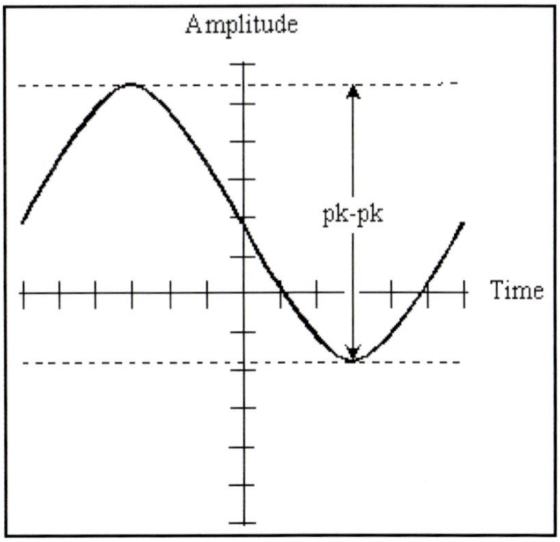

Peak to peak measurement

PARTS LIST: RF DIODE PROBE	
Capacitors:	2 x 0.1µF (100nF) disc ceramic
Diodes:	2 x germanium or silicon small signal diodes
Screened lead	

A germanium diode, like the common 1N34A, is good choice, although I commonly use small silicon signal diodes such as the 1N914 or 1N4148. Silicon power diodes, 1N4001 or similar, are less suit- able for RF work because of their higher internal capacitance.

The probe can be a stiff piece of copper wire to touch the circuit point being tested. A wire with a small crocodile clip provides a ground connection. In practice I rarely make a ground lead but use a clip lead connected between the ground of the circuit under test and the ground of the probe. It is good practice to connect the probe to the meter using a screened lead to prevent the meter picking up any stray RF signal. The lead could be 50-ohm instrument cable, but cheap microphone screened lead will also do the job.

The RF Probe can be built in a variety of ways. **Fig 19.2** shows a layout using the 'island' method of construction described earlier. The three blocks at the top represent island pads and the long block represents the ground connection. In practical terms, this could be three 'islands' mounted over a ground plane.

The picture below shows an RF Probe built using the commercial copper island pads described earlier in the book. The three circular island pads are at the top of the picture and the ground connections go directly to the copper clad ground plan. Note the PVC sleeve pushed over the stiff wire probe. This is culled from a piece of thick insulated wire.

An alternative method is to cut out a matrix of copper squares in a piece of printed circuit board material. This is most easily achieved by drawing a hacksaw blade across the surface of the board until it just cuts through the copper layer. Holding the board firmly in a vice and gently applying the saw blade should produce neat cuts in the copper.

Small pieces of copper wire (excess cut from the capacitor leads) are soldered across

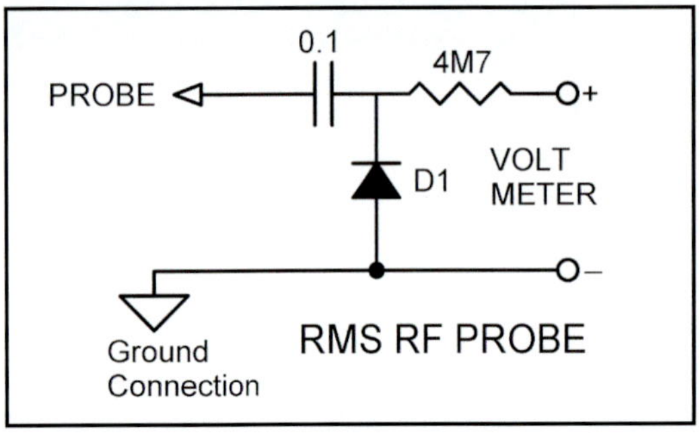

Fig 19.3: Circuit of an RMS radio frequency probe

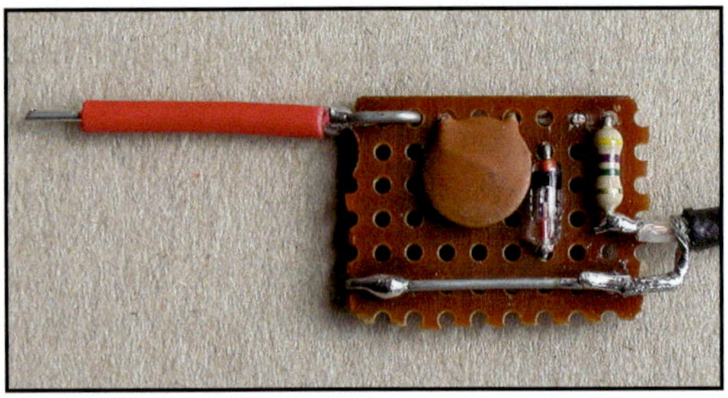

Perfboard RMS RF probe

the bottom three copper squares to make the ground connection. I have also added an extra strip across the top to make the probe easier to hold. This type of RF probe gives the peak-to-peak voltage measurement. The graphic shows what might appear on an oscilloscope display with an RF signal. It is oscillating with positive and negative peaks either side of zero volts. The RF probe circuit used above is a 'half wave voltage doubler' circuit, which gives the total voltage from peak to peak of the signal (pk-pk in the graphic). The 'peak' voltage is half the peak-to-peak voltage (zero to either peak) and the effective power of the signal is even less. These peak values are only transitory during the cycles of oscillation. The effective power is the root mean square value (RMS); that is 0.707 times the peak value.

I tend to build and use peak-to-peak RF Probes for QRP work since they are useful in detecting small RF signals. It is, however, very simple to build an RMS RF Probe that is accurate enough for amateur radio applications.

Fig 19.3 shows a simple RMS RF probe suitable for use with a digital multi-meter. The single diode produces a peak voltage reading

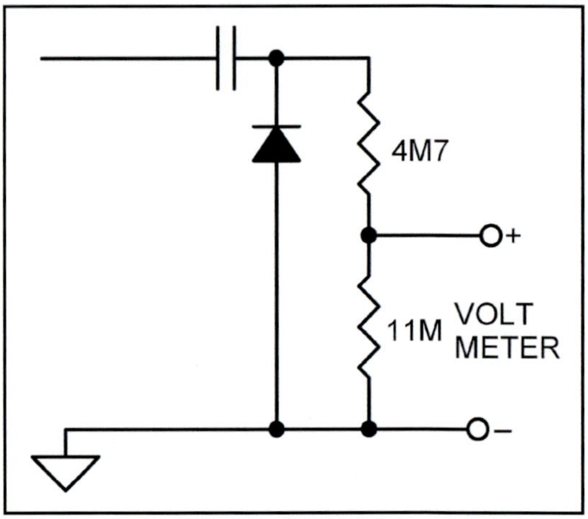

Fig 19.4: How the RMS voltage divider works

and the large value

series resistor roughly performs the calculation required to give an RMS voltage reading. Digital meters have a high input resistance (usually about 10 – 11MΩ).

In effect, what happens in this RMS RF Probe is that the 4.7MΩ resistor forms a resistive divider with the 11MΩ internal resistance of the voltmeter to obtain the RMS value of the peak voltage produced by the diode.

(**Fig 19.4**). Ideally the 4.7MΩ value ought to be adjusted according to the actual resistance offered by the voltmeter. For example, if the resistance of the meter is 10MΩ, the value should be 4.4MΩ. The accuracy is good enough though for most amateur's use. The photograph above shows the simple RMS RF Probe built on a small piece of perfboard. The interconnections are beneath the board.

One of the problems in using a simple RF Probe is picking up stray radio frequency signal. The best precaution against this is to use screened leads, as shown in all the photographs, to shield the probe components from the hands of the user. The body can pick up

PARTS LIST: RMS RF PROBE	
Capacitor:	0.1µF (100nF) Disc Ceramic
Diodes:	1N34A (or similar – see above)
Resistor	4.7M Ω

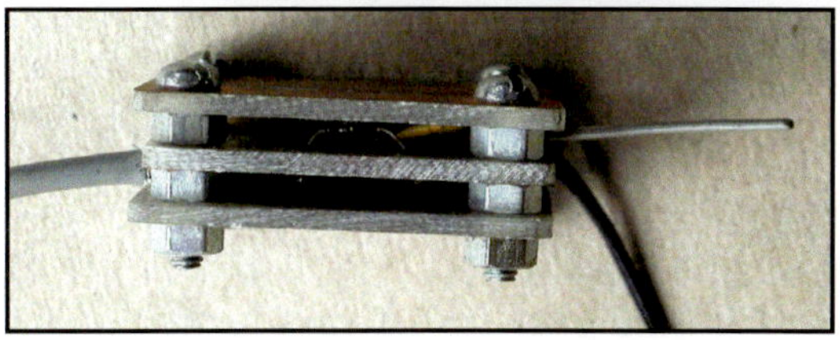

RF probe shielded with print circuit board material

RF signals very effectively, so some builders mount the circuit in a small metal box, but a simple method is to add front and back plates from printed circuit board material.

The two photographs show another RF Probe that I built with two shield plates either side of the circuit board. The plates are made of printed circuit board material, cut to the same size as the probe circuit board and using nuts as spacers. This is my usual workshop probe.

The RF probe is used in conjunction with a voltmeter (usually a voltage range on a multi-meter) to determine the presence and the amount of radio frequency signal. The ground clip is attached to a ground point on the circuit being tested and the probe touched on to the point being tested. The capacitor at the probe isolates the probe from any supply DC voltage present at the test point.

The probes described above assume a low voltage supply. If a probe is to be used on valve equipment with high voltages, the input capacitor should have a voltage rating well above the maximum voltage present in the equipment being tested. *Only an experienced technician should attempt to use a probe on high voltage equipment.*

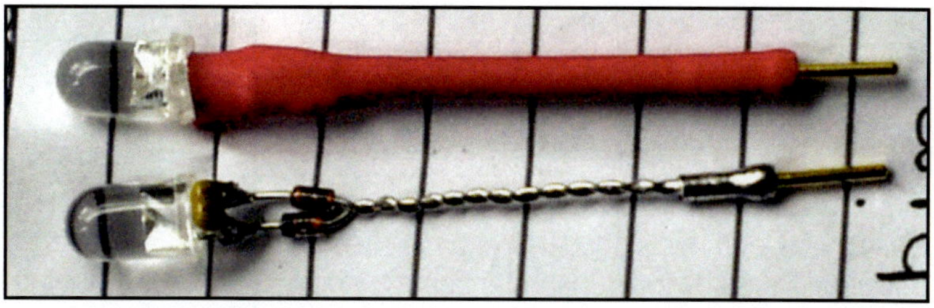

The LED RF sniffer probe (showing construction method)

A simple RF sniffer

Sometimes it is just enough to know that a radio frequency signal is present in a circuit. I am indebted to that fine radio constructor, George Burt, GM3OXX (SK), for a very simple little device for 'sniffing out' a radio frequency signal.

In an item in the RSGB's RadCom ('Technical Topics', May 1997), G4TKV, pointed out that Superbright and Hyperbright LEDs (light emitting diodes) will illuminate at very low levels of cur- rent and are ideal for detecting radio frequency signals. GM3OXX built a simple LED RF test probe using a Hyperbright light emitting diode (**Fig 19.5**).

The probe wire leads to a pair of Schottky barrier diodes; these have a low forward voltage drop and give enough voltage to illuminate the LED at low levels of detected signal. The probe is held with a finger and thumb either side of the LED body to pro- vide a ground plane. The LED will light when held an inch or so from the output of a transmitter as low as about half a watt and will indicate RF present at very low levels by touching the circuit.

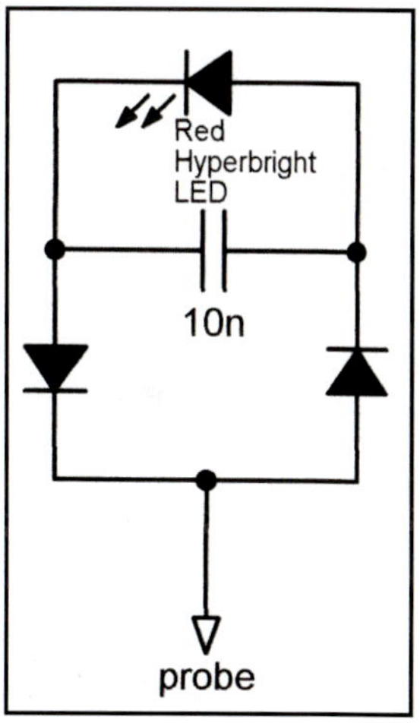

Fig 19.5: Circuit of the LED RF probe

GM3OXX built

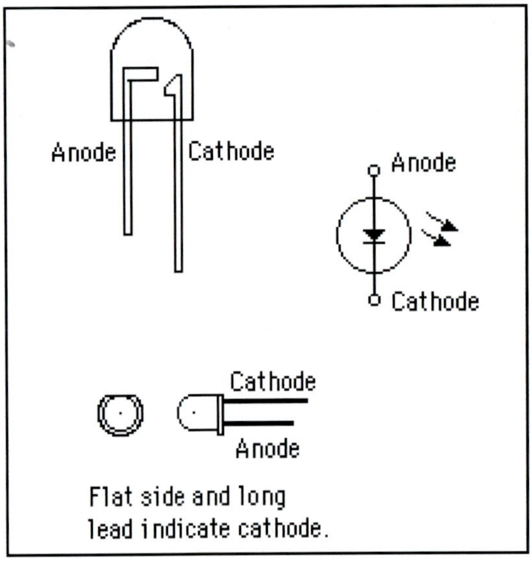

LED pin connections

PARTS LIST: LED RF PROBE	
Capacitor:	0.01iF (10nF) small disc
Diodes:	2 x BAT86 Schottky barrier
LED	Red Hyperbright

his prototype by using a very small capacitor, mounted close to the leads of the LED. The two diodes have short leads attached to the LED and capacitor, and the input leads are twisted together and terminated with a printed board pin to form a probe. The body of the probe is covered with a heat shrink sleeve.

Dummy loads and RF measurement

Another essential little item of test equipment for the transmitter builder is the dummy load. Almost all QRP transmitter circuits are designed with an output impedance of 50 ohms. Many of the designs use power amplifier transistors, which can be damaged unless a suitable load is present on the output. It is good practice never to test a QRP transmitter without such a load to absorb the generated RF power. A dummy load is a 50-ohm resistance that can handle the power delivered by the transmitter being tested.

It is easy to obtain high wattage resistors that are wire wound; that is made from resistance wire wound on an insulated former. These are not suitable for use as a dummy load because of the inductance formed by the winding. Always use carbon or metal oxide based, non-inductive, resistors for making a dummy load. These are not easily available with a high wattage rating so usually a dummy load is made from several resistors. The required 50 ohms is made up by series or parallel combination of resistors, and the resistors share the power. Readers can refer to the Resistors / Capacitors Appendix at the back of this book for how to work out the values of combinations of series and parallel connected resistance.

Fig 19.6 shows a typical arrangement for using paralleled resistors to make a dummy load. A simple example would be to parallel-connect two 100-ohm resistors, or four 200-ohm (220 ohms would be a standard value - near enough). The power sharing would be such that two 100-ohm, 1-watt resistors will handle a power of 2W; and four 200-ohm, 1W resistors will handle a power of 4W.

The photograph shows a dummy load made from standard off-the-shelf quarter watt resistors. Twenty 1kΩ, quarter-watt, resistors are connected in parallel to give a 50-ohm, 5W dummy load. Notice that the resistors are connected in a fan-like structure. Dummy loads are best built without

long leads and avoiding any configuration that might add inductance/capacitance to the circuit. I made this arrangement by taping the resistors round a small piece of wooden dowel. The leads at the top end were bent over the end of the dowel and bound together with a length of solid copper wire. The bundled and bound

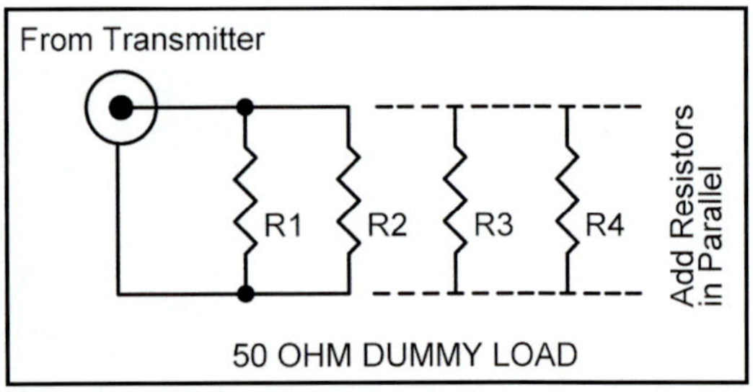

Fig 19.6: A dummy load using paralleled resistors

Dummy load made from low wattage parallel connected resistors

A series connected dummy load

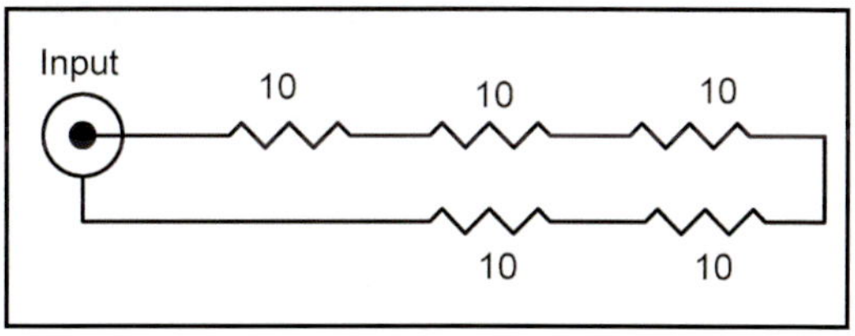

Fig 19.7: A series connected dummy load

wires were soldered and the ends trimmed off. A loop of copper wire was then wound around the other end of the resistors, each one being soldered to the loop and the ends trimmed. After removing the resistors from the dowel, the bundled end was connected to the centre of an RF connector and the loop to the sleeve of the connector.

Fig 19.7 shows a series connected dummy load. This uses five 10-ohm, 1W resistors to give a 50- ohm, 5W load.

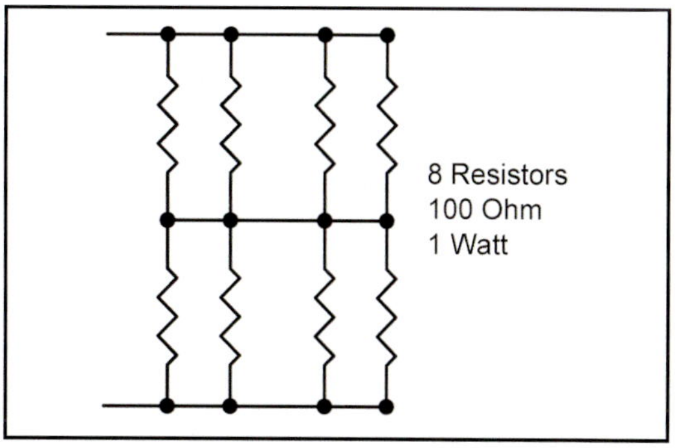

Above right: Fig 19.8: Series/parallel dummy load made from eight 100-ohm resistors

Those who are happy to play around with Ohm's Law and calculate values based upon available resistors may like to use combinations of series and parallel values. **Fig 19.8** shows how eight 100- ohm, 1-watt resistors can be used to make a 50-ohm, 8W dummy load.

Measuring watts

A dummy load and an RF probe provide all the necessary parts for a simple meter to measure the output power of a transmitter.

Fig 19.9 shows the circuit of the RMS RF Probe connected to a dummy load. Assuming the voltmeter is a digital DC voltmeter with an input resistance of about 11MΩ, it will record the RMS voltage across the dummy load. A simple calculation will do the rest.

The equation required is:

$$\textbf{Power} = \mathbf{V_{RMS}2 / 50}$$

In other words, the RMS voltage squared divided by 50, which is the resistance of the load as shown.

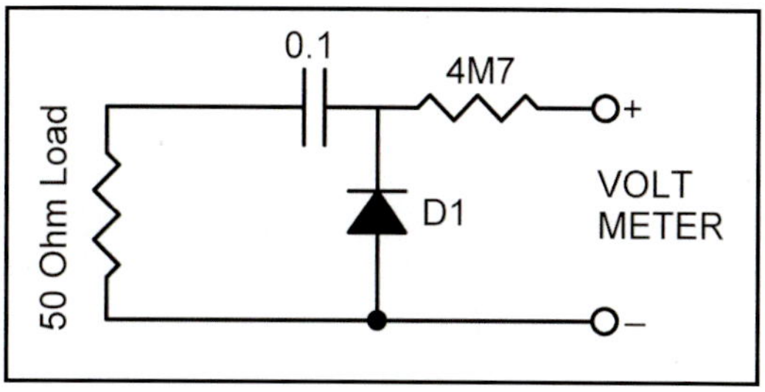

Fig 19.9: Measuring output power

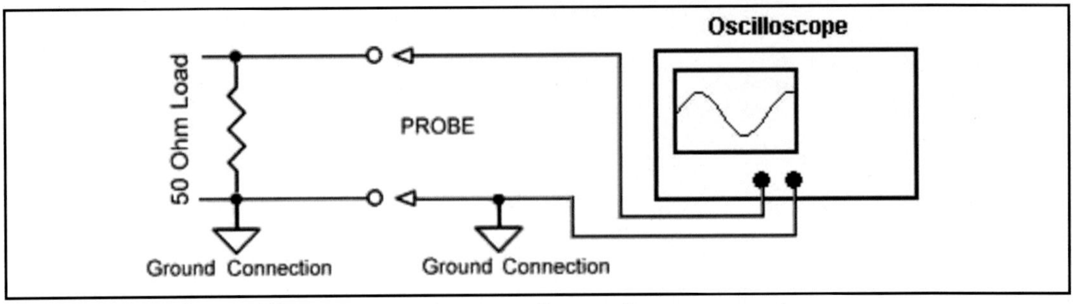

Fig 19.10: Use of an oscilloscope to measure waveform or frequency spectrum

Evaluating signals by waveform and spectrum

In chapter 16 ("The QRP workshop") the oscilloscope/spectrum analyser type of kit now commonly available was mentioned as being very desirable. However, like all good tools, it is useful to know how to use such equipment to the best advantage and to avoid pitfalls for the unwary, in so doing. The oscilloscope, for example, is very helpful in evaluating both repetitive waveforms and also single shot types of waveform but the question is "Is the waveform being displayed actually that present in your circuit?". An example will show what I mean. Supposing we wish to examine the waveform across a 50 ohm resistive load, and this is the output of say, an RF amplifier:

Typically, the input resistance of an oscilloscope is around 1 MW up to 10 MW, but there is usually some associated capacitance with that resistance, usually between 10 pF and 20 pF, which modifies the frequency response of the oscilloscope. For this reason, oscilloscope probes include frequency compensation, which is set up using a test square wave which the oscilloscope itself often generates. This is by way of a compensating capacitance which equalises the overall frequency response from the probes' connections up to the oscilloscope itself. Consider the true connection of an oscilloscope input to the above 50 ohms load, as shown in **Fig 19.11**:

If there was no frequency compensation, as in **Fig 19.11 (a)**, then at high frequencies, and/or higher load impedances than the 50 ohms shown, then the shunt capacitance of 15 pF could lower the voltage actually available for measurement by the oscilloscope. The use of a frequency-compensated potential divider, as shown in **Fig 19.11 (b)**, means that the capacitance presented to the load resistance (or impedance) across which the voltage or spectrum is being measured is no more than 1.5 pF in the example shown and the frequency response is lev-

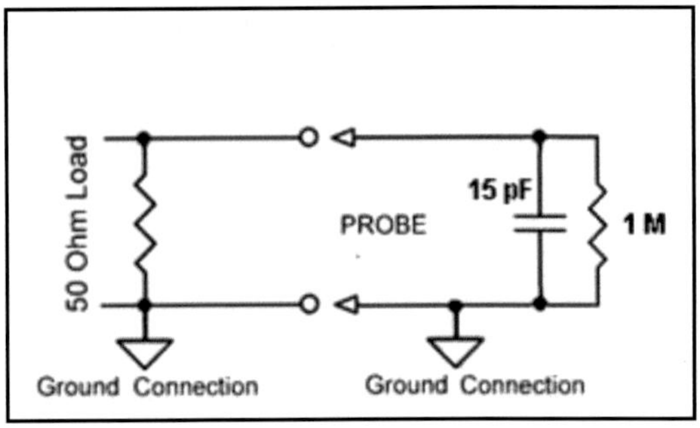

Fig 19.11(a) Oscilloscope measurement without frequency compensation

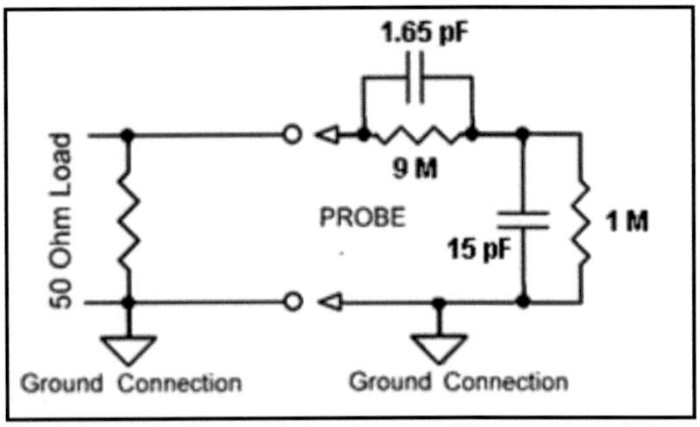

Fig 19.11 (b) Oscilloscope measurement with compensated x10 probe

elled, i.e. flat. However, the price paid is a reduction in sensitivity by a factor of 10. Usually this is not too important as oscilloscope amplifiers are very sensitive. To get exact frequency compensation, the 1.65 pF capacitor is usually a small variable capacitor, of, say 10 pF maximum value, fitted within the probe (as is the 9 MΩ resistor as shown), and adjusted using a small screwdriver to give a square wave on the display when a square wave source is fed into the probe.

Bench setup of M0RJG

The bench setup of the Roger, M0RJG, complete with digital oscilloscope/spectrum analyser in the top left, soldering equipment, copper tape, and home-made, as well as commercial signal generators. The ongoing project in this case was an upconverter for the 10-metre band, using a 4 MHz source mixed with 25 MHz using an SA612 balanced mixer/modulator.

Chapter 20

Practical Projects – Transmitters

When considering your first QRP transmitter, a good starting point is to build a reliable and well-trusted design either from 'fresh ingredients' (buy the parts) or from a kit. The transmitters described below are available in kit form.

The OXO transmitter

The OXO transmitter is one of the classics of the QRP world, so called because it was designed by George Burt, GM3OXX (SK), and it has been available in kit form for many years. It is a crystal-controlled transmitter; that is, a quartz crystal oscillator circuit determines the frequency. Operating a single frequency transmitter may seem to be a severe limitation. However, a stable VFO (variable frequency oscillator) can be difficult for a beginner to build, whereas crystal-controlled transmitters have a low component count and are very likely to give immediate and satisfactory results. Most crystal oscillators can be configured to work as a VXO (variable crystal oscillator). A VXO transmitter with a crystal on one of the QRP calling frequencies should pull in two-way QRP contacts. Suitable crystals may be obtained from some of those listed in the traders appendix.

The OXO transmitter contains only three stages: a crystal oscillator, a power amplifier and a switching transistor. For greater simplicity, the switching transistor, which keys the power amplifier, could be omitted, although this is not really advisable as the transmitter would have to be keyed by switching the power supply directly with the Morse key. It can be made to work on any amateur band from 80m to 20m with the addition of a suitable crystal and low-pass filter.

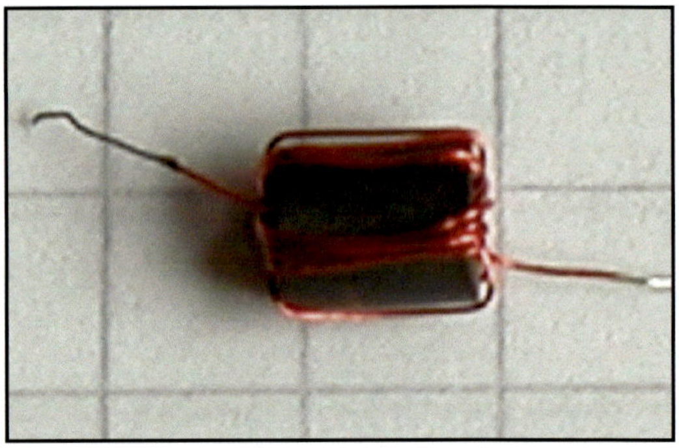

Detail of the radio frequency choke

A health warning!

Building a modest transmitter is a good way to try QRP operation. Many seasoned QRPers began by building a small transmitter from a kit or a circuit design. But beware! Once that first little transmitter has been built, probably in one evening with little financial outlay, it may put the commercial HF station in jeopardy. It is easy to get so involved in the satisfaction of contacting other stations on diminutive homemade transmitters that the expensive commercial transceiver may get pushed to the back of the bench to gather dust. Some radio amateurs never recover!

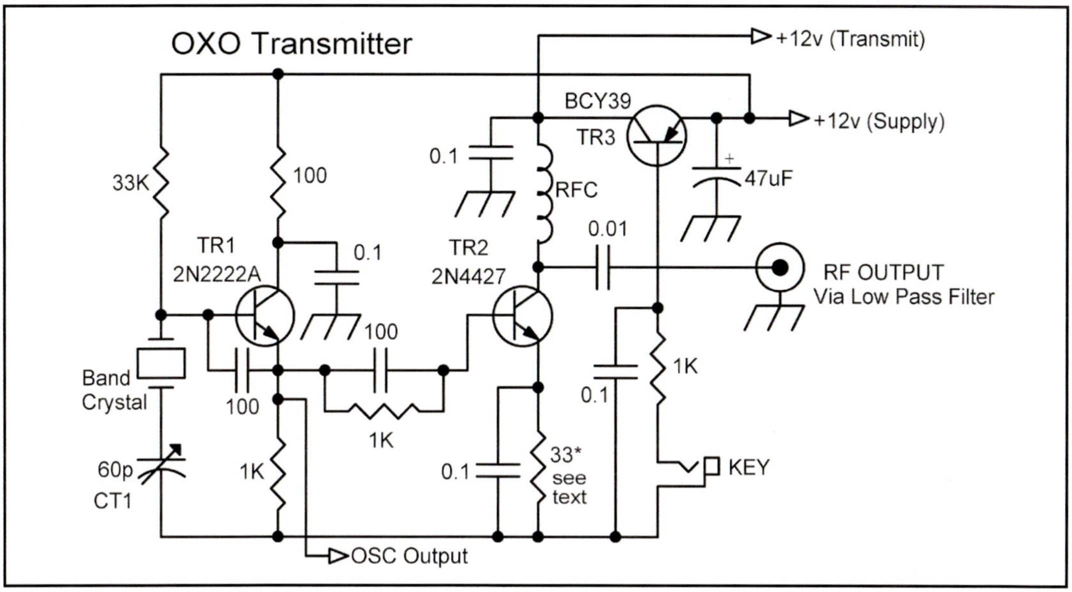

Fig 20.1: The OXO transmitter

Let us take a look at the transmitter circuit (**Fig 20.1**). TR1 is the crystal oscillator, the 100pF capacitor provides feedback signal from the emitter to the base of the transistor to maintain oscillation. The variable capacitor, CT1, allows some change in the frequency of the oscillator – a simple form of VXO. If a fixed frequency transmitter is required, the bottom of the crystal may be connected directly to ground. The oscillator signal appears across the 1kΩ resistor in the emitter of TR1 and is coupled to the power amplifier, TR2, via a 100pF capacitor. The oscillator is also coupled to the power amplifier by a 1kΩ resistor to provide a suitable bias voltage. Fig 20.1 shows a possible oscillator output point should an oscillator signal be required to drive a receiver.

The power amplifier, TR2, has a small value resistor in the emitter circuit, with a 0.1μF capacitor in parallel. The value of this resistor can be adjusted to control the output power of TR2. It is possible to over-drive the power amplifier and damage TR2. The constructor should aim for a power output in the order of 2W. TR2 must be fitted with a clip-on heatsink to keep the transistor cool. If TR2 becomes excessively hot, the value of the emitter resistor ought to be increased. The given value of 33 ohms is a good compromise in most cases.

The output from the power amplifier appears across a radio frequency choke (RFC).

PARTS LIST: WOODEN OXO TRANSMITTER	
Resistors:	33k, 100R, 3 x 1kΩ, 33R [see text]
Capacitors:	4 x 0.1μF [100nF], 2 x 100pF, 0.01μF [10nF], 47μF electrolytic (25V) CT1: 60pF variable or trimmer capacitor
RFC	Radio frequency choke made from 12 turns of 32SWG enamelled copper wire wound through a ferrite bead
Transistors:	TR1 – BC107, BC182, 2N2222A or similar
	TR2 – 2N3866, 2N4427 or similar
	TR3 – BCY70, ZTX751, 2N2905A or similar PNP type
Quartz Crystal	Fundamental frequency to suit the band.

The OXO transmitter built 'ugly style' on copper board

This is a homemade choke wound on a small ferrite bead by passing 12 turns of thin gauge enamelled copper wire through the hole of the bead. This can be a delicate operation requiring careful threading through the hole of the ferrite bead avoiding scratching off the enamel on the edges of the bead. A 0.01μF capacitor takes the output of the power amplifier to a low-pass filter to suit the band.

The transmitter could be keyed by adding a Morse key between the top of the radio frequency choke and a 12-volt supply. This is rather clumsy way to key the transmitter as the supply voltage is across the key contacts; usually one side of a key is connected to ground. To allow this, a switching

transistor, TR3, is added to the circuit. This is a PNP transistor. Any silicon PNP transistor capable of handling the current drawn by the power amplifier would do the job. The 12V supply is applied to the emitter of TR3 and as the base is connected to the ground by the key, via the 1kΩ resistor, 12V appears at the collector. The power amplifier only receives 12V when the key is closed.

The OXO transmitter can be built in a variety of ways. It would not be difficult to build your own version using one of the non-printed circuit board techniques described earlier. The photograph shows an OXO transmitter built using the ugly, point-to-point technique. The jack socket is for the key, and the phono socket is for the output. Notice the clip-on heatsink placed over the power amplifier transistor. Close to the crystal socket is a small trimmer capacitor. The oscillator transistor and the switching transistor are both mounted 'dead-bug' style with the pins pointing upwards.

Most of the components can be mounted any way round. The exceptions are the transistors and the 47μF electrolytic capacitor. Depending upon which transistors are used, find out from the transistor data sheet which leads are the emitter, base and collector. Take special care if a transistor is to be mounted leads-upward. The electrolytic capacitors are polarised and must be mounted the correct way around. This is usually clearly marked on the case.

The Wooden OXO: A little transmitter anyone can build

This version of the GM3OXX OXO classic QRP transmitter is very simple to build. It is capable of working on any band from 80m to 20m. It requires a fundamental crystal for the required band and a low pass filter for

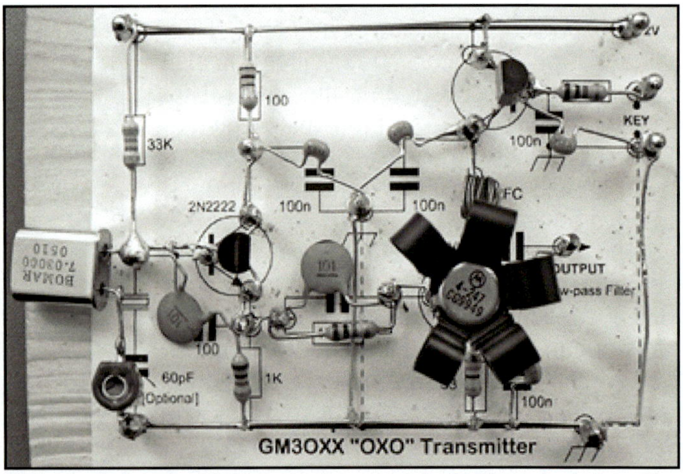

This very simple QRP transmitter can be built to look just like the circuit diagram

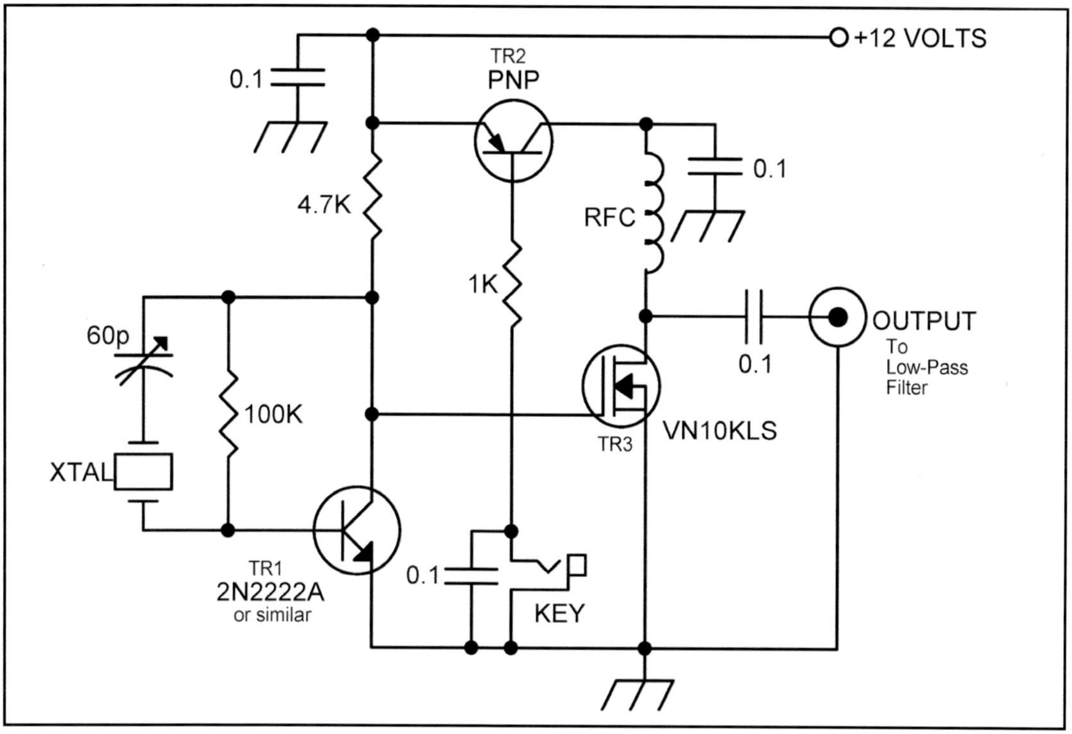

Fig 20.2: The schematic suggests the best placement for the nails and can be photocopied for gluing on to the wooden base

that band attached to the output. It is best run at about 1-watt RF output. The output can be varied by altering the value of R5 but attempts to squeeze too much power from the OXO could blow the power amplifier.

The Wooden OXO is ideal for a beginner because it uses copper plated nails mounted over a copy of the circuit (schematic) drawing, so the parts simply fit over the places shown on the diagram (**Fig 20.2**).

The 40m band is ideal for the OXO and crystals on 7.030MHz, the 'QRP gathering frequency", are readily available

T1 is a crystal oscillator. C1 provides feedback although on some bands the circuit will oscillate without C1. A variable capacitor, VC1, allows the crystal frequency to be pulled a little to produce a VXO (variable crystal oscillator). The variable capacitor is not provided but any capacitor or trimmer around 60pF would do the job. Leaving out VC1 and connecting the crystal direct to ground gives a fixed frequency transmitter.

T1 is RF and DC coupled to T2, so T1 will not oscillate without T2, R5 and C7. The RF load for T2 is a hand-wound choke made from 10 turns wound through a ferrite bead. Scrape the enamel from the wire at each end and tin the bare copper with solder.

T3 is a DC switch, included so that the power amplifier can be keyed with respect to ground. It would be possible to leave out T3 and replace it with a hand key. Note that T3 is a PNP transistor and the emitter lead goes to the 12 volt supply. Power amplifier T2 requires a clip-on heat-sink

Building the OXO is very simple because each part is placed above its position on the schematic drawing as shown in Fig 20.2 and in the photo.

Locating a nail at every component position would result in too many nails on the board and make construction difficult, so I reduced the number of nails.

PARTS LIST:
WOODEN OXO TRANSMITTER

Capacitors:	C1	100pF [101]
	C2	100nF [104]
	C3	100nF [104]
	C4	100pF [101]
	C5	100nF [104]
	C6	10nF [103]
	C7	100nF [104]
Resistors:	R1	33KΩ[orange-orange-orange]
	R2	100Ω [brown-black-brown]
	R3	1KΩ [brown-black-red]
	R4	1KΩ [brown-black-red]
	R5	33Ω [orange-orange-black]
	R6	1KΩ [brown-black-red]
Transistors:	T1	2N2222
	T2	2N3866
	T3	2N3906

Fig 20.2 suggests the best placement for the nails and also shows two added wires (dotted lines) for grounding some parts and two bold lines for the wires used for the +12V and ground.

Carefully hammer the nails about half-way into the wood. Before beginning to add the parts, tin each nail with a hot soldering iron and some solder. Quite a large iron is needed to make the nails hot enough for the solder to flow and to add the parts. The best way to get a good solder joint is to wrap the component lead around the nail, apply heat to the nail and lead and melt the solder on the joint (not on the soldering iron bit). Do not test the transmitter without attaching a 50-ohm load to the output.

Suitable low-pass filters (from the W3N-QN(SK) design) for the amateur bands can be found elsewhere in this book.

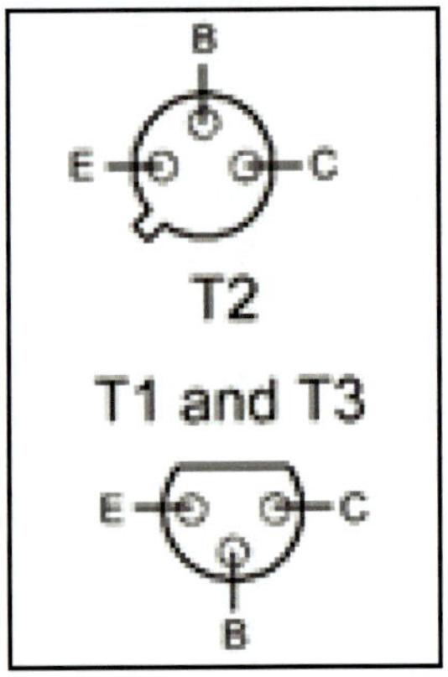

Tansistor pinouts shown from underside

RFC 10 turns through a ferrite bead

The ONER transmitter

Any discussion of simple crystal-controlled transmitters would be incomplete without at least a mention of the ONER transmitter. Alongside the OXO, the two designs are probably the most popular and commonly built basic QRP transmitters. Both have been built in the hundreds and both were designed by GM3OXX. The ONER is so called because it was designed to fit on a one-inch square printed circuit board.

Building a transmitter on a one-inch square board may sound rather daunting but, as the picture shows, the spacing of the parts is sufficient even for a beginner to build the board. The

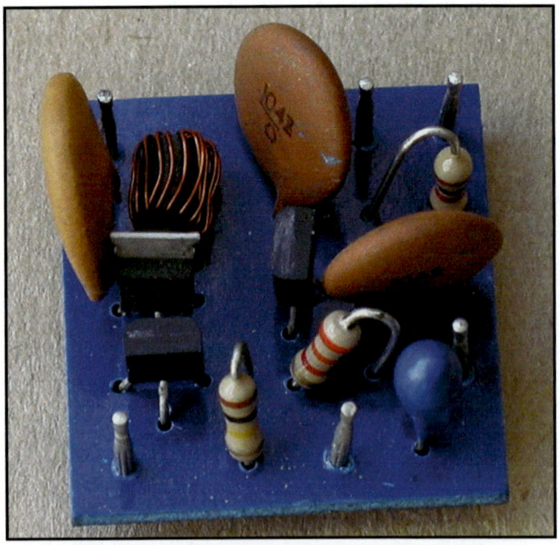
The original ONER kit printed circuit board

ONER is a simple circuit that will lend itself to being built 'ugly style' or using one of the pad construction methods, though such methods are unlikely to produce a one-inch ONER.

The circuit diagram of the ONER is shown in **Fig 20.3**. The parts list overleaf assumes that the ONER may be built without the kit and names alternative transistors.

TR1 is a VXO with a crystal in a feedback path between the collector and the base. TR1 is directly coupled to the power amplifier TR3, which is a MOSFET device. The resistor in the collector of TR1 (4.7kΩ) is the collector load of TR1 but it also provides a bias voltage for TR3. It is possible to reduce the value of this resistor to squeeze more power out of TR3 but too small a value will damage TR3. The value of 4.7kΩ is a very safe compromise, although I have often used 3.3kΩ for this resistor. A home wound radio frequency choke (RFC), as in the OXO, provides the output load for TR3.

The ONER is keyed in the same way as the OXO, using a PNP transistor (TR2) switch. Also, like the OXO, the ONER requires a low-pass filter on the output to reduce harmonic output (see else- where in this book for a suitable filter).

The VN10KLS MOSFET will operate at about the 2W level without the need for an extra heat sink. It has a small cooling tab on the top of the casing.

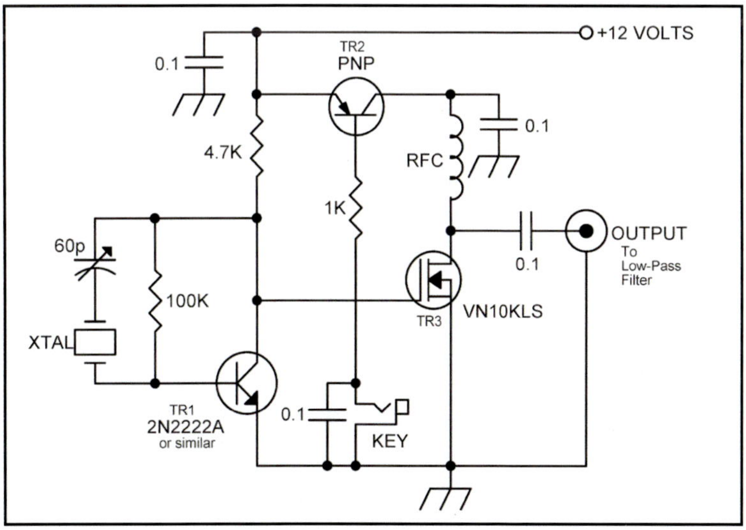

Fig 20.3: The ONER transmitter circuit

PARTS LIST: ONER TRANSMITTER	
Resistors:	100kΩ, 4.7kΩ, 1kΩ
Capacitors:	4 x 0.1µF [100nF]
	60pF variable or trimmer capacitor
RFC	Radio frequency choke made from 12 turns of 32SWG enamelled copper wire wound through a ferrite bead
Transistors:	TR1 – BC107, BC182, 2N2222A or similar (ZTX651 used in the original version)
	TR2 – BCY70, ZTX751, 2N2905A or similar PNP type (ZTX751 used in the original version)
	TR3 – VN10KLS MOSFET
Quartz Crystal	Fundamental frequency to suit the band.

The SP-1 battery powered transmitter

The SP-1 transmitter was designed by George Burt, GM3OXX, as a simple and inexpensive way to put out a QRP signal on the 40m (7MHz) band. The idea for the SP-1 came to George as he waited at a supermarket checkout and saw a display for batteries: "What about a simple transmitter that uses dry battery power?"

The idea was to make it so simple and cheap that anyone could build it and all the parts could be easily found. The choice of transistors for the trans- mitter was very easy as the 2N3904 and 2N3906 types are very cheap. The ones in the prototype cost 6 pence each. Hence the name Six Pence - SP-1. The 2N3904 and 2N3906 are complimentary transistors; The 2N3904 being an NPN type and the 2N3906 being a PNP type.

Data sheets showed that the 2N3906 could handle a current of 200mA and the 2N3904 could only handle 100mA, so the 2N3906 was used for the power amplifier stage (T2). Using a PNP transistor as a power amplifier is a little unusual and results in a circuit that looks 'upside-down' when compared with the other simple transmitter circuits in this book.

T3 is a transistor switch to key the power amplifier. The circuit could be made even simpler by omitting T3 and using a hand-key between L2 and ground, but the use of T3 enables the SP-1 to be used with automatic keyers and gives a little more power out as T3 acts as a series resistor.

A home-wound ferrite bed type choke is used for L2 because it is not easy to find small radio frequency chokes that can handle 200mA. L2 is 10 turns is thin gauge wire (32 or 34SWG) wound through a ferrite bead.

George designed the SP-1 for a 9-volt PP3 battery supply. Using the PP3 the SP-1 shown in the picture gave about 1 watt of RF at 7.03MHz but at this power the 2N3906 was handling about 400mA and getting very warm. Note that a heatsink has been clipped onto the transistor. The small TO-18 heatsink does fit on either the metal or plastic version of the 2N3906.

Using a 6 volt supply the transmitter gave 500mW (half a watt) drawing 300mA through a much cooler transistor. Although these figures will vary with individual transistors, and can differ from manufacturer to manufacturer, perhaps operation below 9V is safer for the amplifier transistor, although at six pence each.

The parts for the SP-1 are commonly available from many component traders. C5 is a 100pF trimmer capacitor which allows some frequency variation of the crystal oscillator. X1 is a 7.030MHz crystal (the QRP calling frequency for 40 metres) available from the GQRP Club and some parts traders. L1 is an axial radio frequency choke; a pre-wound inductor that looks like a fat resistor.

The circuit diagram (**Fig 20.4**) shows a very simple transmit/receive switching arrangement using a

George Burt, GM3OXX is well-known for his simple QRP transmitter designs. The 'ONER' (built on a 1" square board) and the 'OXO' have been built and used by literally hundreds. He only ever uses home-built equipment and never uses more than one watt of transmitter power. In spite of this, he has contacted well over 300 countries on the amateur bands

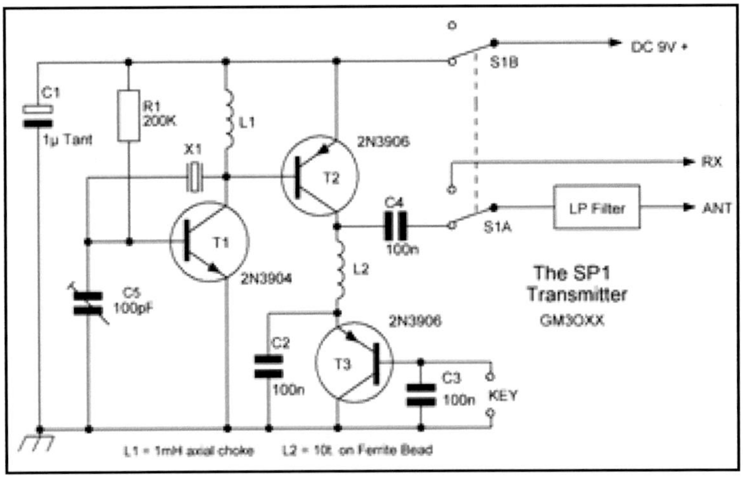

Fig 20.4: Circuit diagram of the SP-1

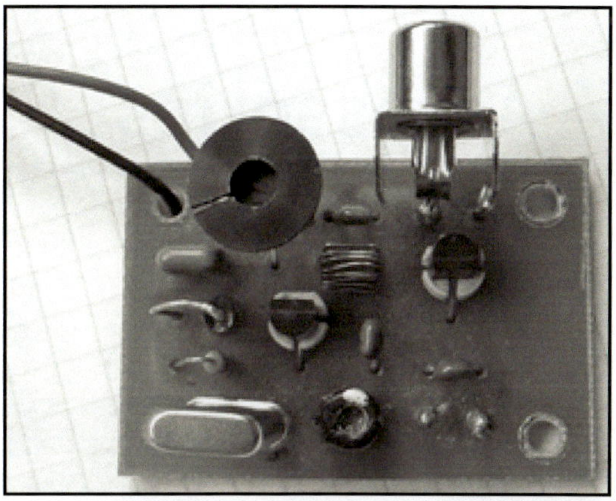

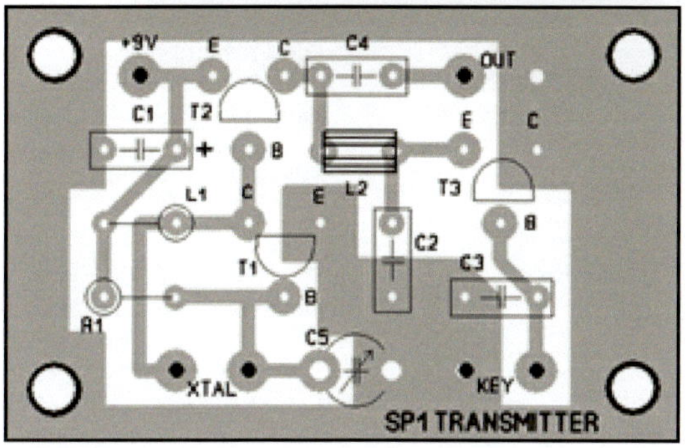

The prototype SP-1 is built on a printed circuit board but the circuit is so simple it could be built using any of the non-printed circuit board techniques described elsewhere in this book. The layout is shown for those who may wish to make their own printed circuit board (NB: Check the size before making a PCB from this picture).

double-pole changeover switch to power the transmitter and to switch the antenna (ANT) between the transmitter output and receiver (RX).

A low-pass filter is required on the output of the transmitter. Even at low power levels, amateur radio stations are required to reduce unwanted harmonic output. The series of W3NQN low-pass filters, described elsewhere in the book, provides an ideal filter for the SP-1.

The SP-1 is a simple way to try out QRP operation. In fact, it is a good test because the power is considerably less than most QRP stations would use. The maximum CW power that can be called QRP is 5W output. The SP-1 will produce less than 1W RF output. In spite of this it is capable of gaining contacts on the 40-metre band used with an available receiver.

The chart below shows what can be done with simple equipment. Ken Evans, W4DU, who lives near Atlanta in Georgia, USA, went on the air during a major international DX contest using an SP1 transmitter in conjunction with the simple Sudden receiver described elsewhere in this book. Between 1423 and 1505 on 22 February 2009, Ken had three amazing contacts for such a simple set- up. To be able to contact Italy, Anguilla and the Madeira Islands from the south of the USA using very simple equipment is surprising . . . except to those who use QRP!

CHART OF W4DU CONTACTS WITH SP-1 TRANSMITTER AND SUDDEN RECEIVER

Callsign	Band	Power	UTC	Country	Freq
IR2C	40m	750mw	02:23	Italy	7.033
VP2E	40m	750mw	02:42	Anguilla	7.032
CT9L	40m	750mw	03:05	Madeira Is.	7.030

Chapter 21

Practical Projects – Accessories

A simple SWR bridge

In the section on setting up a simple station, I described the use of an SWR (Standing Wave Ratio) bridge to use between a transmitter and an antenna tuner. Some antenna tuners have an integrated SWR bridge. At QRP levels of power, they are simple to build.

Fig 21.1 shows the circuit diagram of a 'resistive' SWR bridge. The bridge is not only simple to build but it also offers some protection to the transmitter power amplifier. A resistive load is offered to the transmitter output when the bridge is being used so even if there is a severe mismatch between the transmitter and the tuner, the power amplifier is protected.

The principle is very much like the Wheatstone Bridge, so beloved of physics masters when I was at school. The antenna forms part of the bridge and when it is matched to 50 ohms, the meter reads zero.

Since the bridge adds resistance to the path of the signal it will reduce the output, therefore the bridge has to be switched out of the output path once the measurements have been made. In the circuit a double-pole change-over switch brings the bridge in and out of the signal path.

The photograph shows the SWR bridge built on perfboard. The resistors for R1, R2 and R3 are made up from two 100-ohm resistors

PARTS LIST: SIMPLE SWR BRIDGE		
Resistors	R1	2, 3 – 50 ohm, 2W or greater [could be paralleled resistors]
	R4	1kΩ
	R5	10kΩ potentiometer [or pre-set]
Capacitors:	C1	2 – 0.01iF [10nF] disc capacitor
Diode	D1	Germanium type 1N34A or OA91 etc. or small silicon type 1N914 etc.
Switch:	S1a/b	Double-pole change-over switch
Meter	Moving coil	Full scale deflection 1mA or less

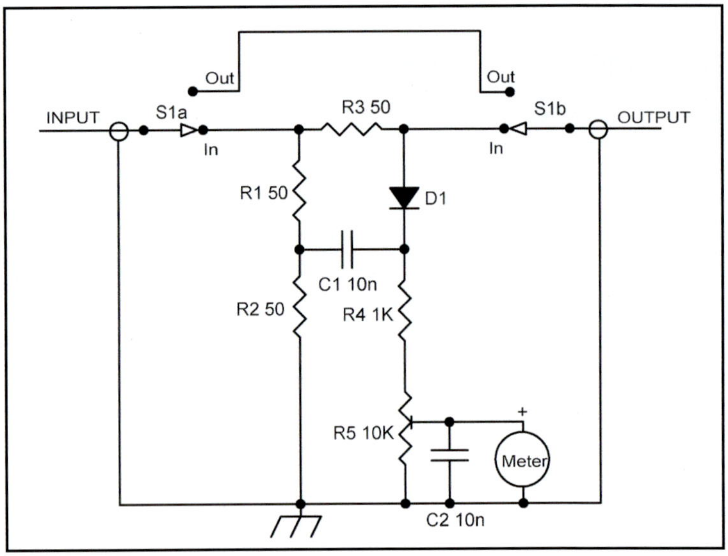

Fig 21.1: A resistive SWR bridge

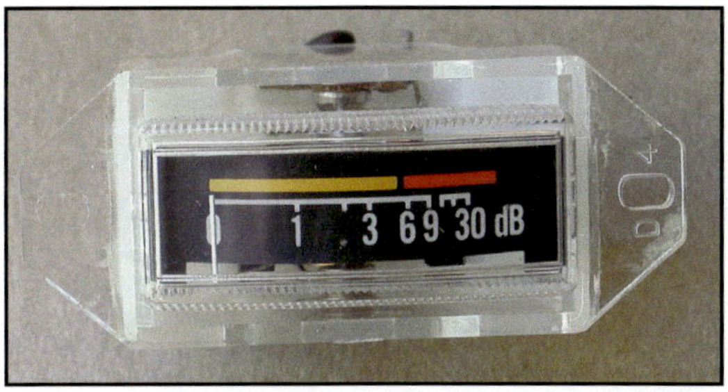

Above: CB type S-meter suitable for the SWR bridge

in parallel. R5 is a pre-set resistor, which can be adjusted to give an adequate reading. The diode in this example is a 1N914 silicon signal diode, but germanium diodes like the 1N34A or OA91 are better because of the smaller voltage drop. The meter

can be any moving coil type that gives a full-scale deflection of 1mA or less. The cheap ex-CB-radio edgewise reading S-meters are ideal, typically they have a full-scale deflection of around 200μA.

The SWR bridge is inserted in the signal path between the transmitter (or transceiver) and the antenna tuner. It remains switched out for normal operation of the transmitter. When tuning up, the bridge is switched in to the path of the signal and the antenna tuner is adjusted for the minimum reading. The setting of R5 is adjusted to give a high enough reading to allow measurements to be made.

By adding a two-pole, three-way wafer switch, as shown in **Fig 21.2**, it is possible to set up the meter according to the level of the output signal. In the 'set' position, R5 can be adjusted to give a full-scale deflection on the meter. Position 'ref' reads the reflected power in the signal path and 'out' switches the bridge out of the signal path.

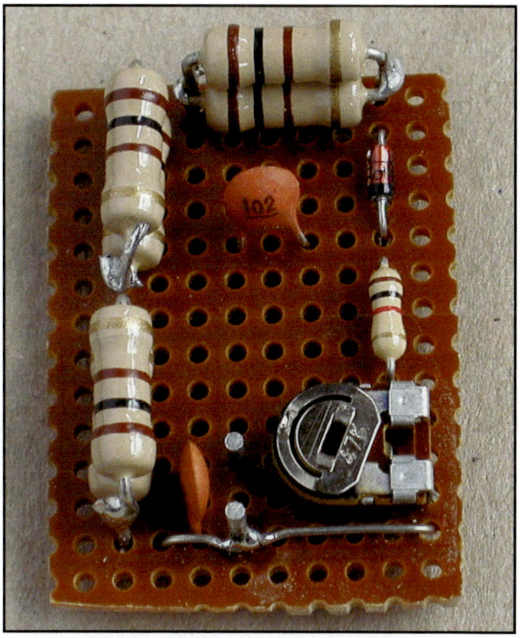

Right: SWR bridge built on perfboard

Simplest SWR bridge

The same resistive bridge principle can be used to make a very simple SWR Bridge using an LED in place of the meter. The property of the Superbright and Hyperbright LED to illuminate at low current levels is used in the circuit shown in **Fig 21.3**. An LED is the null indicator in a bridge formed by 50-ohm resistors and the output load. A switch chooses between the bridge and a direct path for the signal. Using the bridge is simple. When antenna tuner adjustments are to be made, the bridge is switched in to the signal path. The tuner is adjusted for minimum illumination of the LED. The bridge is switched out for normal use of the transmitter.

The photograph shows the bridge built on a small piece of printed circuit board material. Parallel resistors are used to give the appropriate power rating for the 50-ohm resistors. The switch is a double-pole, double throw, toggle switch. The LED

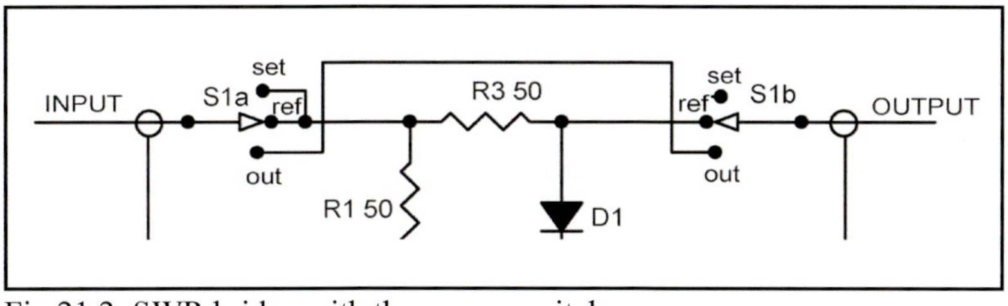

Fig 21.2: SWR bridge with three-way switch

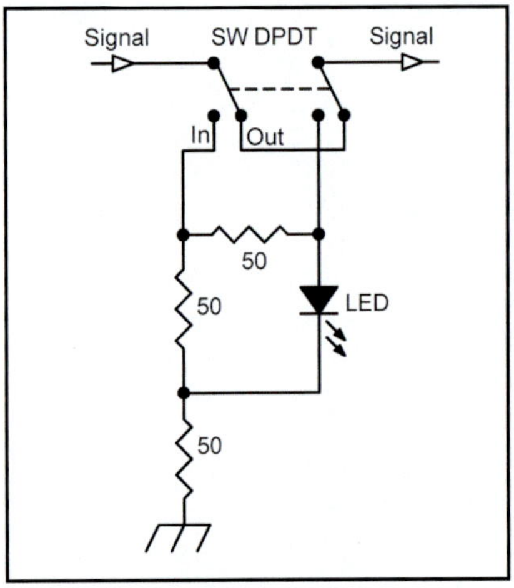

Fig 21.3: Resistive SWR bridge

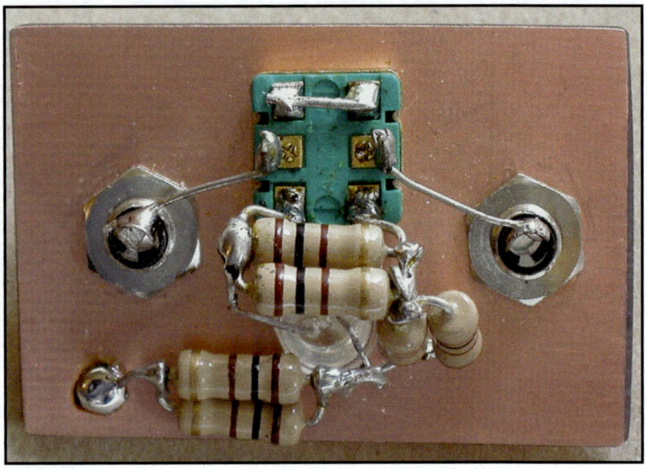

LED SWR bridge built ugly style with LED

PARTS LIST: SIMPLEST SWR BRIDGE	
Resistors	3 x 50 ohm (2W or greater), each may be made from several resistors
LED	Superbright or Hyperbright
Switch	Double-pole, double-throw

(a 'jumbo' Hyperbright LED) and the input and output sockets are mounted in holes drilled in the printed circuit board material.

The Stockton bi-directional in-line wattmeter

David Stockton, GM4ZNQ, designed one of the most popular measurement devices used by QRP operators, the Stockton bi-directional in-line wattmeter. It uses a hybrid transformer arrangement (**Fig 21.4**) to measure the forward and reflected power in a trans- mission line. As David says himself: "The operation of this circuit is difficult to understand, yet it only uses two components. Fortunately, it is easy to build and easy to use". Essentially two simple transformers are used to sample the for- ward and reflected RF power in a 50-ohm coaxial feeder. It can be used as a power meter and for matching a transmitter output to an antenna. The wattmeter remains in the signal path the whole time and does not impede the signal. It is a true in-line device.

The transformers, T1 and T2, are wound on T50-43 ferrite cores. In the original design ferrite cores made by SEI were used, but these are difficult to obtain. The main winding is 12 turns of PVC covered solid copper wire occupying about three-quarters of the core.

The coupling (shown as a single turn on the diagram) is about 2 inches (5cm) of 50-ohm coaxial cable mount- ed between two insulated standoff pillars and passing through the centre of the cores. The centre and screen of the cable are trimmed and connected as shown in the layout drawing. Solder tags are used for the ground connections.

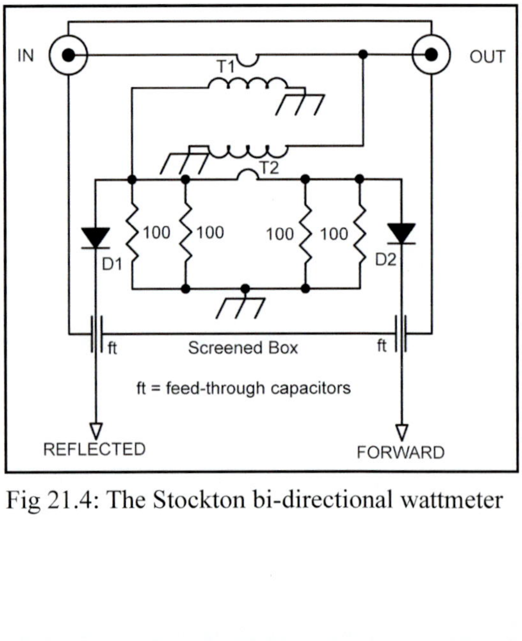

Fig 21.4: The Stockton bi-directional wattmeter

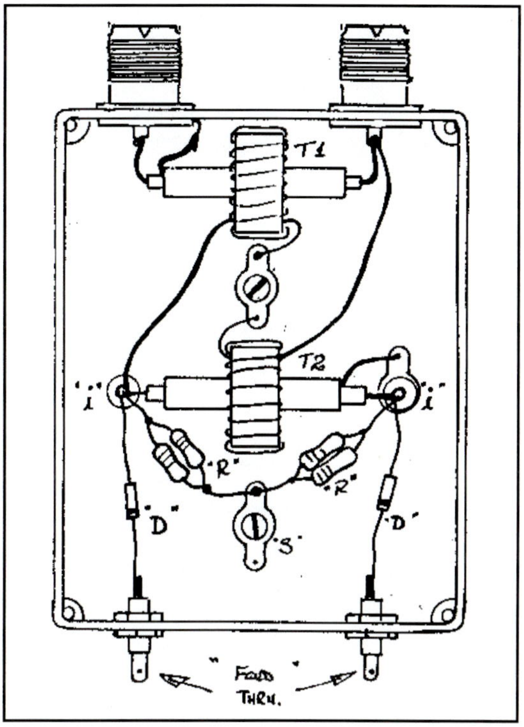

Fig 21. 5 Stockton bi- directional wattmeter layout

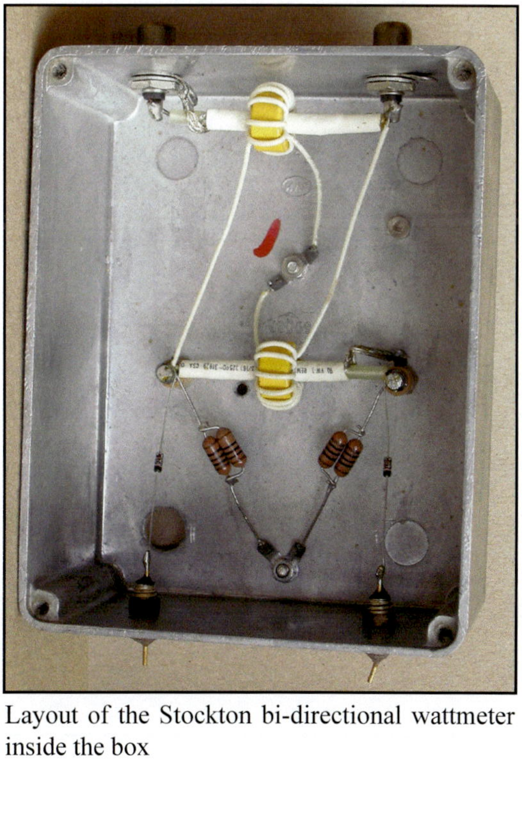

Layout of the Stockton bi-directional wattmeter inside the box

The output from the diodes leaves the screened box via two feed-through capacitors.

In **Fig 21.5**, 'R' are the resistors, 'i' are the standoff pillars, 'S' is a ground solder tag. T1 is mounted directly across the RF connectors for the input and output and T2 is mounted on the standoff pillars. The short lengths of coaxial cable passing through the centre of the cores are only grounded at one end; at the socket for T1 and at a grounded solder tag for T2. The photograph shows a practical example of the bi-directional wattmeter built in an aluminium die-cast box.

The Stockton bi-directional wattmeter requires two meters (**Fig 21.6**) to indicate the forward and reflected power. In the prototype 50μA moving coil meters were used with series resistors to adjust the full-scale reading.

The series resistors could be pre-set resistors to adjust the full-scale reading. As a rough guide, 22kΩ would be suitable for 5W full scale and 56kΩ for 20W full scale. Both meters could be calibrated in watts by using the built meter in series with an already calibrated wattmeter. The dummy load, RMS probe and digital voltmeter combination described previously could pro- vide a method of calibration.

The completed unit has the distinct advantage of continuous monitoring of forward and reverse power levels. It can remain in the signal path between the transmitter output and the antenna

PARTS LIST: STOCKTON BI-DIRECTIONAL WATTMETER

Resistors	4 x 100-ohm low inductive resistor (1 or 2W)
Capacitors	2 x 0.01µF (100nF) or 0.01µF (10nF) feed-through capacitors
Inductors	T1 and T2 - 12 turns PVC covered, single strand wire (see text)
Diodes:	D1, 2 – Schottky Diodes (BAT81, BAT82 or similar)
RF Sockets and ground tags	
Metal enclosure	

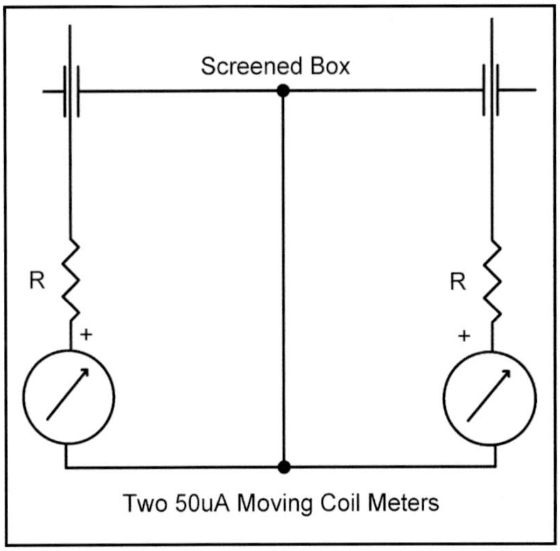

Fig 21.6: Adding meters to the Stockton bi-directional wattmeter

Original SM7UCZ Coupler

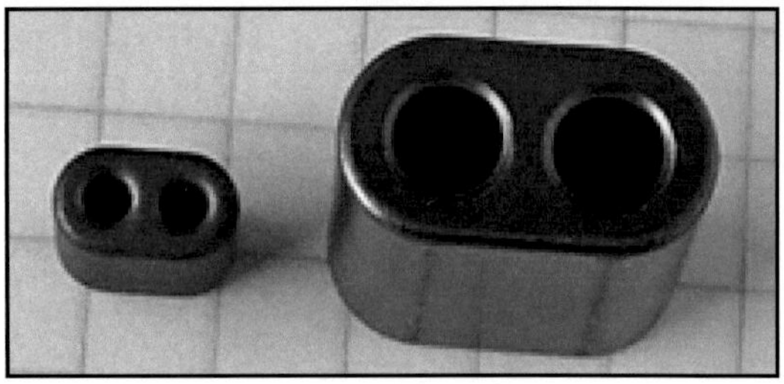

Binocular Cores

tuner, and can be used to monitor both the output power and the matching between the transmitter and the antenna.

Alternative Stockton bi-directional coupler

Johnny Apell, SM7UCZ, has suggested an alternative to the toroid core transformers (T1 and T2) in the original Stockton design. SM7UCZ sent me a photograph of his version using a single ferrite binocular core to make the coupler.

Binocular cores, sometimes called 'pig nose cores' (look at the shape!) have two holes passing through the ferrite material. They are available in several sizes. The type designated as BN-43-202 works well in this circuit and its size aids in winding the coils.

The circuit (**Fig 21.7**) follows the original coupler configuration except that each 12-

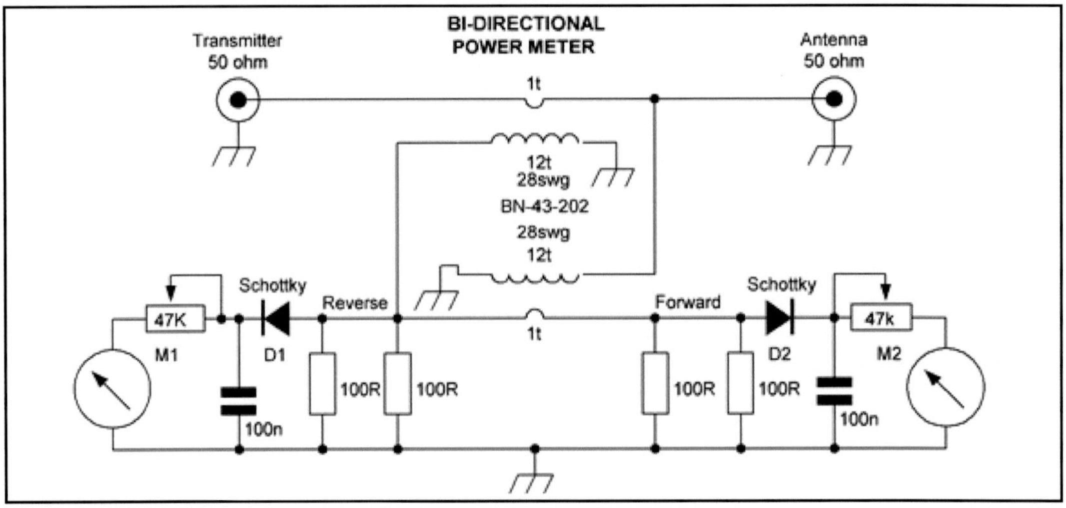

Fig 21.7: Alternative Stockton circuit devised by SM7UCZ

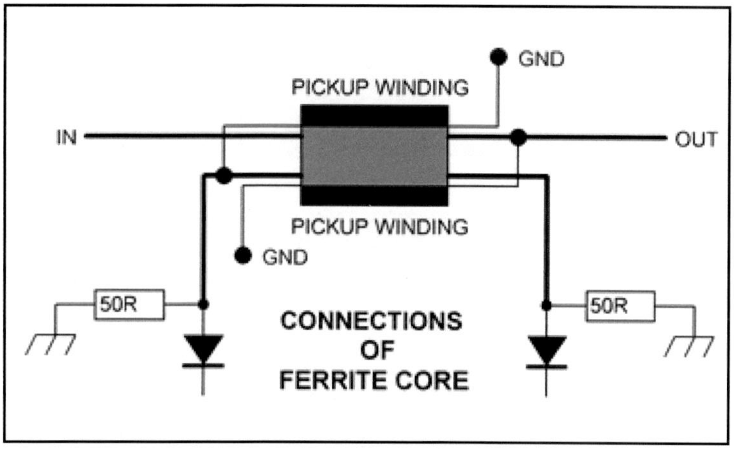

Fig 21.8: Connections around the ferrite core

A prototype of the alternative Stockton bi-directional wattmeter

turn coupler coil is wound on the two holes going through the BN-43-202 binocular core. The signal passes through the coupler from left to right. The forward voltage appears at the right-hand side 50-ohm load and the reverse (reflected) voltage appears at the left-hand 50-ohm load. Note that two 100-ohm resistors in parallel have been used to form the 50-ohm load resistors. The load resistors should be able to handle enough power to cope with the power of the signal; the two parallel resistors share the total power.

My prototype was built, ugly style, on a piece of printed circuit board material. The photograph shows the physical arrangement of the coupler. A 12-turn winding of 28SWG enamelled copper wire is wound through each hole of the binocular core.

A single wire carrying the signal passes directly through the top hole, and the sensing winding is another single wire passing through the bot- tom hole of the core. Making the coupling trans- former is easy but does require a little care. Each time the wire passes through the hole counts as one turn. Pull the wire to be against the side of the hole as each turn is made but avoid scratching off the enamel coating on the edges of the hole. As far as possible make the windings lie side by side. The single turn (single wire) through each hole is made with fairly stiff wire; I used 22SWG tinned copper wire. The stiffness of these wires holds the coupler in place as shown in the photograph. The coupler is suspended between the input and output sockets. I used phono sockets. Individual constructors can use input and output sockets to match their equipment.

Wiring the pickup windings in the correct place is critical. Thankfully this is easy because the physical connections match the coupler layouts in the circuit diagram. **Fig 21.8** should make this clear.

The rest of the circuit is built 'Manhattan style' using connection pads glued to the copper clad board. The forward and reverse voltages appear across the two 50-ohm resistors and these radio frequency voltages need to be converted to direct current voltages. Ideally a pair of Schottky diodes (such as the BAT82 or 1N5711) should be used but germanium diodes, or even the common silicon diodes would do the job, albeit with slightly reduced output. The pre-set resistors and meters provide the voltage measurement. Surplus 'VU type' edgewise reading meters with a full-scale deflection in the order of 100 - 200 μA can be used. The pre-set resistors are set according to the power levels being used. This circuit is only suitable for powers of 10 watts or less.

An RF powered changeover system

Earlier I described how a method of changing the station from transmit to receive is required. This could be a manual switch, to switch the antenna to either the transmitter or the receiver. If the transmitter is powered during the receive periods, the transmitter signals will be heard in the receiver and interfere with the required signals, so the oscillator must also be disabled. For a crystal-controlled transmitter this can be achieved by only powering the transmitter during the transmit periods as. for all practical purposes, a crystal-controlled transmitter locks on to the correct frequency quickly when switched on.

A more convenient way to perform the changeover operation is a system which eliminates the use of an extra transmit-receive switch. There are several designs that use the key to operate an electronic switch to move the antenna between the receiver and transmit- ter. A system that I have often used is powered from the RF signal generated by the transmitter.

The operation of the RF changeover circuit (**Fig 21.9**) is quite simple. When the relay is not energised, one side of the normally closed contacts connect the antenna (ANT) to the receiver (RX) and the normally open contacts disconnect the antenna from the transmitter. If RF power comes from the transmitter, the other side of the normally closed contacts pass that power to the 50-ohm resistance.

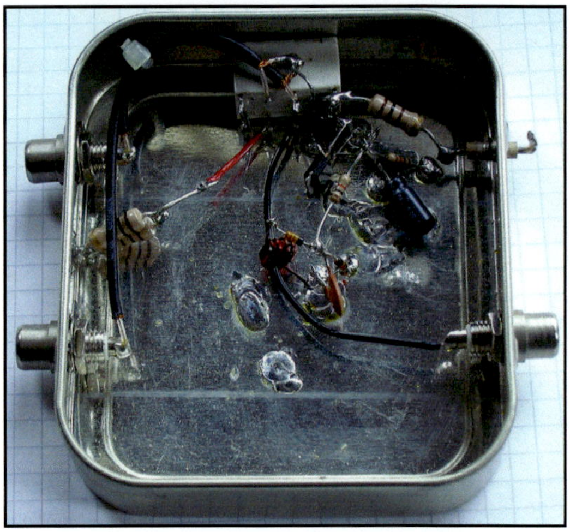

The changeover circuit built 'ugly style'

The wire carrying the RF power passes through the centre of a ferrite toroid. The toroid has a pick-up winding - five turns of a suitable gauge wire. This coil picks up some of the RF signal, which passes to a pair of diodes acting as a detector. The DC voltage from the diodes switch on the

VN10KLM MOSFET and the relay is energised, causing it to switch over the contacts. The antenna is switched from the receiver to the transmitter.

As the transmitter is keyed on and off, the relay would also switch on and off, causing the contacts to clatter on and off. To prevent this, a delay circuit is placed in the gate of the VN10KLS. The 10kΩ resistor and 47μF capacitor hold the MOSFET on for a short period of time. These values have been chosen so that the relay remains switched on at normal keying speeds. When there is a pause in the keying, the relay switches off, bringing the antenna back over to the receiver.

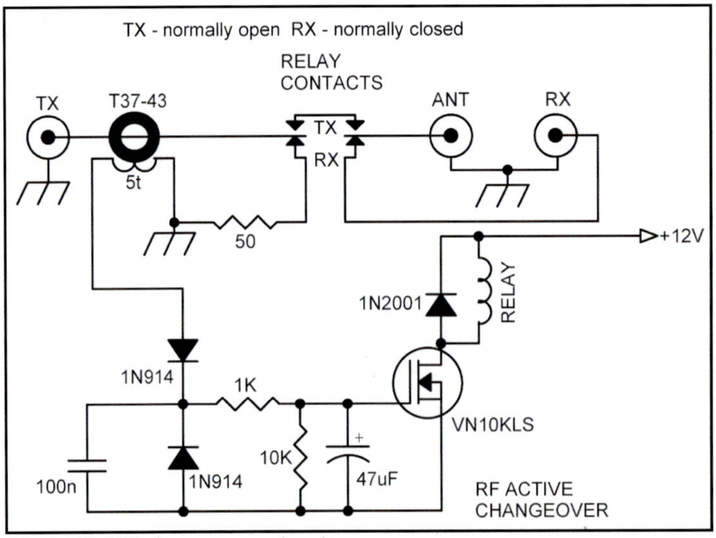

Fig 21.9: RF changeover circuit

PARTS LIST: RF POWERED CHANGEOVER SYSTEM	
Resistors:	50 ohms (to suit power of transmitter) 10kΩ
Capacitors:	0.1µF (100nF), 47F electrolytic (16V wkg)
Semiconductors:	2x 1N914 diodes or similar
	1N2001 diode
	VN10KLS mosfet
Relay:	12V with double pole changeover contacts
T37-43 ferrite core	
3 x RF sockets (to suit equipment)	

I have one of these units connected permanently to the back of my station stand-by receiver. It is plugged in to the receiver and the antenna so that I can connect and use a little crystal-controlled transmitter at any time. My example (pictured) is built, very much ugly style, in a small pipe tobacco tin.

Relay controlled transmit-receive changeover

An alternative to the RF powered changeover is to have the transmit/receive control activated by the Morse key. This is a key actuated change-over system that can be used alongside any little QRP transmitter. The only pre-requirement is that the keying of the transmitter should be in respect to ground. The circuit diagram is shown in **Fig 21.10**. The transmitter key controls two DC switching transistors; T1 and T2. A diode isolates this switching circuitry from the transmitter circuitry. T1 and T2 are PNP transistors. Any suitable devices capable of handling the current required from the '12V keyed' line would be suitable. 2N4036 and BCY71 transistors are suitable, as are many other PNP types.

T1 operates as a DC switch to provide a '12V Keyed' supply. This facility is an optional extra in that the change-over function could be built without T1 and its associated circuitry. Having a 12-volt source that comes on and off in sympathy with the keying action can be very useful. The illumination of the LED will follow the keying action.

Transistor T2, another PNP device, is also switched on and off by the action of the key. T2 in its turn operates an NPN switching transistor, T3, to drive a relay for the change-over function. A whole variety of common NPN silicon transistors would work for T3; the limitation being that T3 must be able to handle enough current to operate the relay. I used a BFY51 but also any common PNP transistor would do the job.

T2 actuates T3 via a time delay circuit controlled by the 330-ohm resistor, the 10KΩ trim-pot and the 100uF electrolytic capacitor. When the key is closed, T2 switches on T3 and the relay closes. On releasing the key, the relay is held in for a time governed by the decay of the voltage across the 100uF capacitor. The decay time may be adjusted by the 10KΩ trim-pot. Having this delay ensures that the relay does not clatter in and out in sympathy with the keying action. The trim-pot can be set so that the relay only falls out when there is a significant pause in the keying action. Individual users will be able to set this according to speed of their keying.

The two diodes in the emitter of T3 hold the emitter voltage up to about 1.4 volts as a safeguard against any leakage current in the transistor holding in the relay once T3 is switched off. The diode in parallel with the relay coil acts as a voltage spike suppressor when the electromagnetic field in the coil collapses.

All diodes are shown as 1N914 silicon diodes but almost any silicon diode would serve in any of the diode locations. The relay can be any small relay capable of reliable hold-in with 12 volts DC and having two sets of change-over contacts. The relay contacts switch the antenna between the receiver and the transmitter. The second set of contacts are used to provide 12 volts on transmit and 12 volts on receive. The circuit diagram shows the relay contacts set in the 'receive' position, with the relay unenergised.

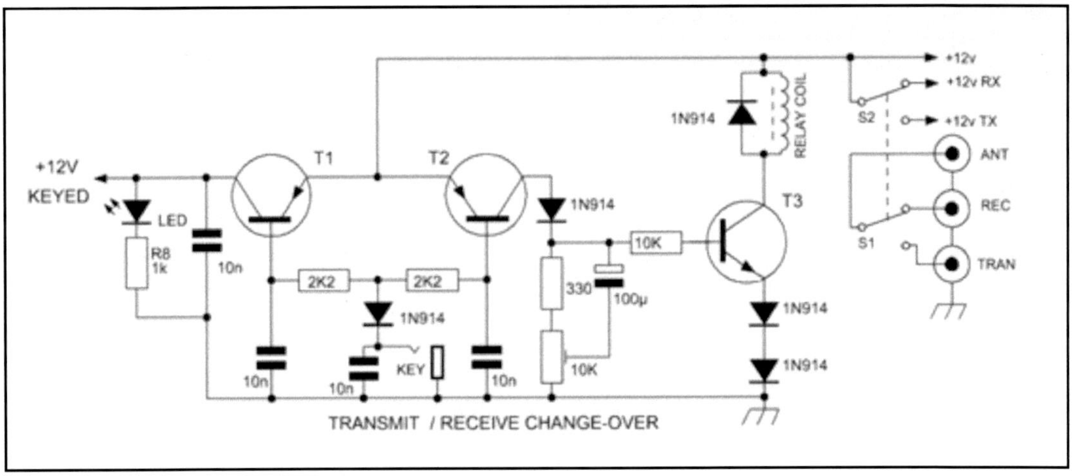

Fig 21.10: Circuit of the relay-controlled T/R changeover

A twin-T audio oscillator

Many CW operators find it difficult to key a transmitter without hearing what they are sending with the Morse key. It is usual for transceivers to have a system to listen to the keyed signal.

This is either done by listening to the actual transmission through a heavily muted receive section or having a separate audio oscillator operated by the key. Such an oscillator is usually called a 'side tone generator'.

A very basic QRP station, made up from a crystal-con- trolled transmitter and a receiver, need not include a side tone generator. The simplest, and crudest, method is just to turn down the receiver volume during the transmissions and listen to the transmitter signal. This will overload the receiver, not enough to cause damage but enough to produce an unpleasant sounding signal. So, some operators turn the volume control right down to zero or even devise a system to completely mute the receiver. In this case, a side tone oscillator will be needed to monitor the keying.

Many simple transmitters, like the ONER and the OXO, use a transistor DC switch between the key and the transmitter. This produces a 12-volt output that follows the action of the key. A convenient way to add a side tone oscillator is to take the supply from this 12-volt keyed point on the transmitter. This will switch the audio oscillator on and off in time with the keying action.

Almost any audio oscillator circuit will serve as a side tone oscillator - it just has to produce a sound in time with the keying. I have commonly seen simple multi-vibrator circuits used as side tone oscillators but these produce a harsh, rasping, sound because of their square-wave output. The simple circuit I prefer to use is the twin-T oscillator (**Fig 21.11**). The twin-T uses two resistor-capacitor feedback

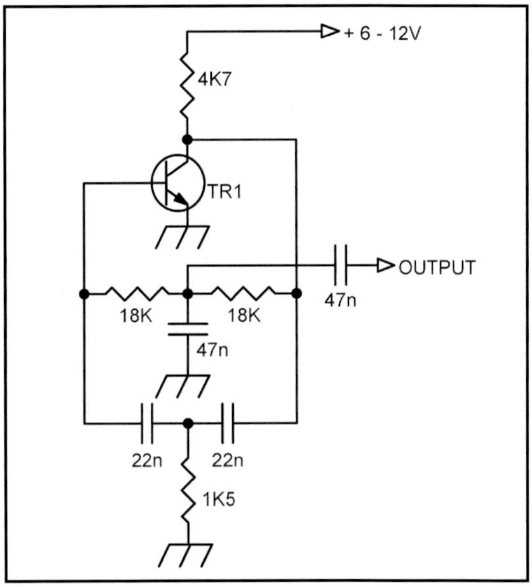

Fig 21.11: Circuit of the twin-T audio oscillator

PARTS LIST: TWIN-T AUDIO OSCILLATOR	
Resistors:	2 x 18KΩ, 1k5, 4k7
Capacitors:	2 x 22nF, 2 x 47nF (disc ceramic)
Transistors:	TR1 – NPN - 2N2222A, BC182, etc.

networks to produce a sine wave output that gives a much sweeter note. It uses very few parts and is ideal for ugly construction on a piece of copper clad board.

The Twin-T oscillator is simple to build using any of the methods described earlier. The supply may be taken from a 12V keyed point on the transmitter. The output is low but is enough to directly drive a crystal earpiece or even a crystal microphone insert used as a sounder.

Alternatively, the output could be fed to an audio amplifier or to the audio stages of a receiver. The twin-T audio oscillator would also make a good oscillator for Morse code practice.

Polarity protection

This is an important addition to any project. When bench testing, it is so easy to connect the power supply the wrong way around and damage circuit components. Polarity protection is so simple and worth including in any home-built equipment. Four methods are shown below.

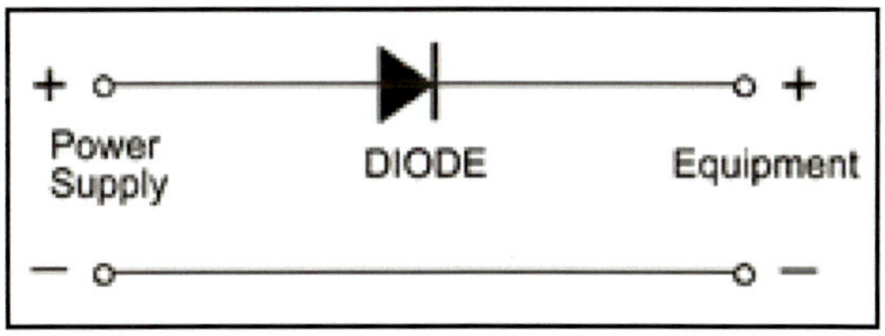

Fig 21.12: Simple diode protection

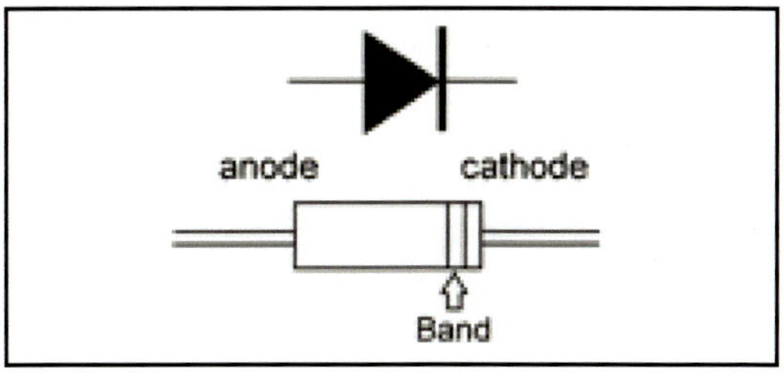

Fig 21.13: Diode polarity and markings

Fig 21.12 shows the simplest system; putting a diode in series with the power line. This is some- times called the 'idiot diode'! As all readers will know a diode only allows current to pass one way; it is a form of valve (see **Fig 21.13**). Connecting the diode as shown in the positive line means that current will flow only when this line is connected to the positive side of the power supply.

The disadvantage of this system is that there is a small voltage drop across the diode; in the order of 0.7V. This may not be critical for the circuit but has to be considered.

The arrangement shown in **Fig 21.14** is absolute fail-safe. A diode bridge is connect- ed in line with the power supply. The bridge is such that whatever way round the power supply is connect- ed, the correct polarity will be applied to the equipment. This circuit is very safe but suffers from the same problem as the simple diode protection circuit (Fig 21.12), only more so. The supply passes through two diodes on the way to the equip- ment and therefore more than a volt of the supply voltage is lost.

The circuit of **Fig 21.15** is yet another approach. A diode is connected across the supply as shown. If the polarity of the supply is wrong, this diode will short the supply to ground causing the fuse to blow. This ap- proach does rely on the fuse blowing before any damage is cause to the equipment. A fast-blow fuse should do this successfully but there is a chance it may not stop the damage in time.

The solution shown in **Fig 21.16** is my favourite which has proved to be a very reli- able protection although it is more complex than the others. The positive side of the power supply passes through a relay switch that only switches on the supply if the polarity is correct.

The positive line is connected to the relay switch and the equipment is attached to the normally open contact (NO) of that switch.

This means that until the relay is energized, the supply does not reach the equipment. A 12-volt relay is required as the coil is energized by the supply line.

The diode, D1, is connected so that only a positive voltage will energise the relay coil. If the supply line is at positive D1 will pass current to allow the relay switch to close and power to reach the equipment. Diode D2 is added to reduce voltage spikes in the relay coil. I can commend this circuit to any equipment builder.

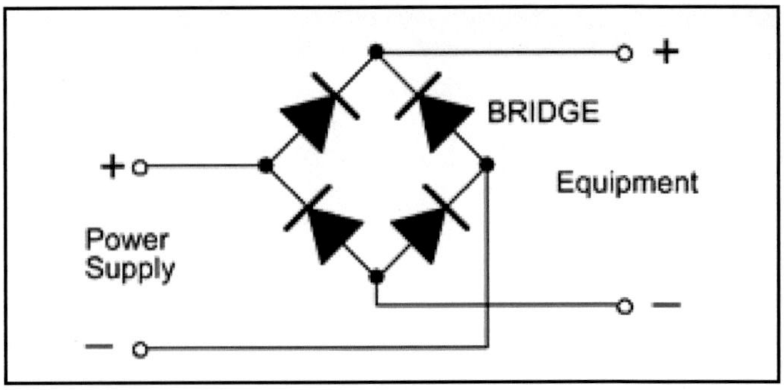

Fig 21.14: The bridge circuit produces a fail- safe solution, but at the expense of voltage drop

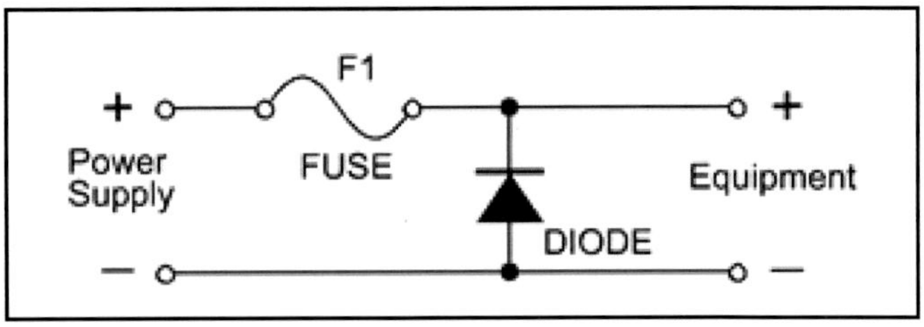

Fig 21.15: A fused protection circuit

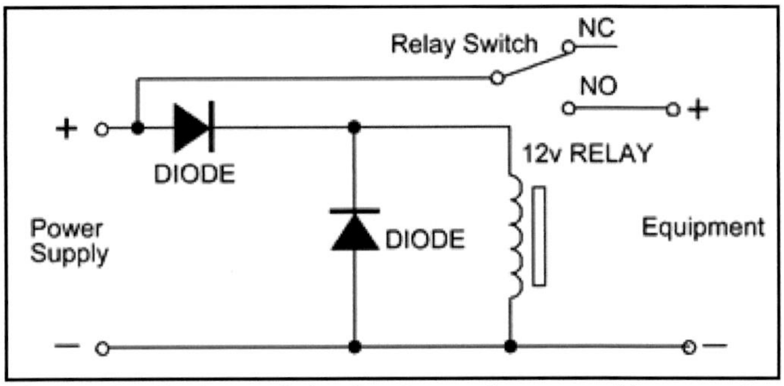

Fig 21.16: A fool- proof circuit using a relay to provide protection

Chapter 22

Three Projects for Beginners

These three simple projects for the beginner comprise a transmitter and receiver for the 40 metre (7MHz) band and a Z-match antenna tuner. These designs began life as 'Buildathon' projects. A Buildathon is where a group of beginners build the same project together under the guidance of more experienced radio constructors. All were featured in Buildathons at the GQRP Club convention in Rishworth, West Yorkshire, England.

For ease of building, with opportunity to correct any errors, the 'Manhattan' style of construction was used (see the earlier chapter on construction tips and techniques). Manhattan construction was so named by Chuck Adams, K7QO, who thought that the parts standing above the base board looked like a Manhattan skyline. It is a technique familiar to Australians as 'Paddy Board Construction'.

W4DU describes the Manhattan method as a development of the so-called ugly construction where components are soldered to each other above a piece of copper clad board. The copper board forms a ground plane and parts that require grounding are soldered directly to the copper base board.

The Manhattan style introduces small rectangular or circular copper pads with other special pads that are glued to the copper board, but are insulated from the ground plane.

These provide convenient places to connect components to each other and maintain isolation from the ground. Integrated circuit chips require a 'carrier' that has the footprint of the IC pins (shown earlier in this book). These projects can be built using homemade or commercial pads and IC carriers.

The Buildathon 40m transmitter

The Buildathon Transmitter (**Fig 22.1**) is about the simplest circuit possible for a reliable QRP transmitter; a single crystal-controlled oscilla-

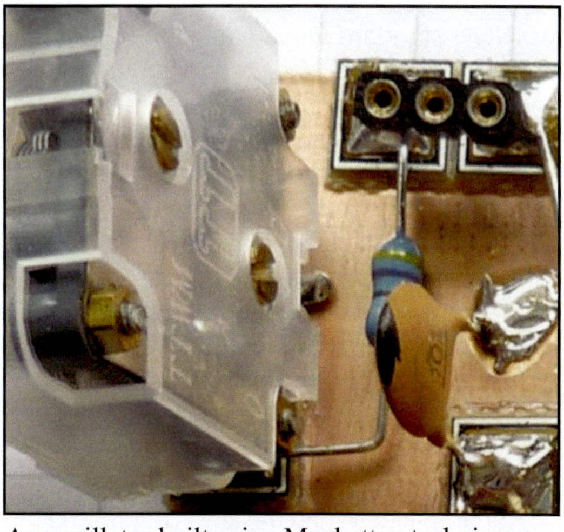

An oscillator built using Manhattan techniques.

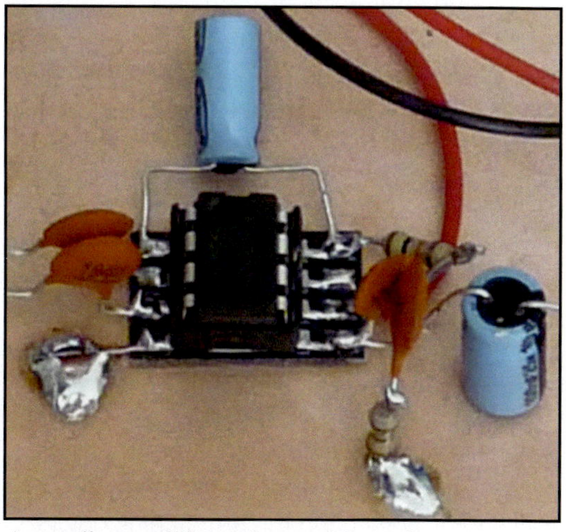

An audio amplifier using an IC carrier.

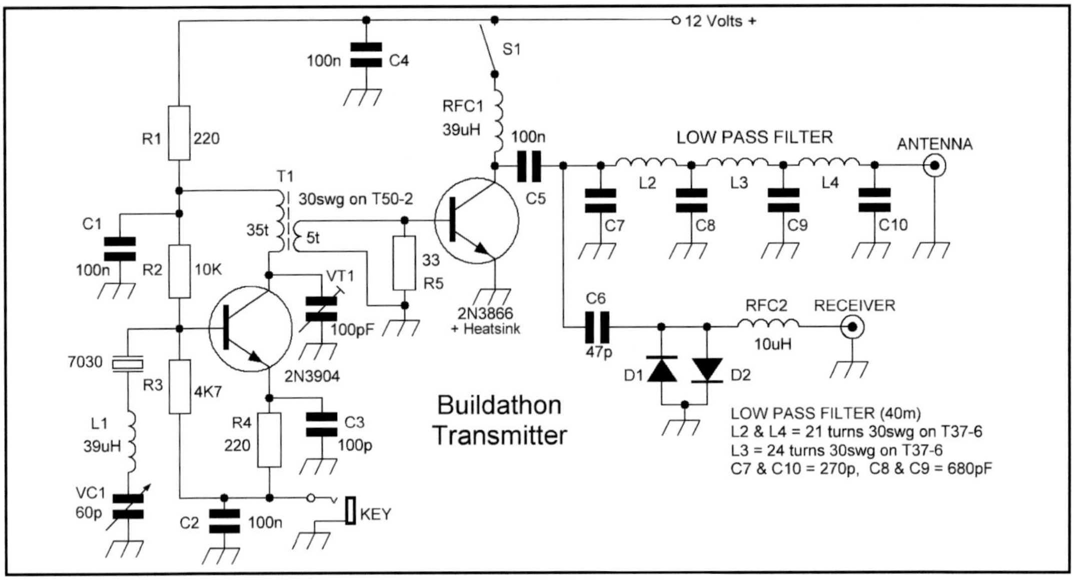

Fig 22.1: Circuit of the Buildathon 40 metre transmitter

tor drives a single stage power amplifier. It is based on the W7ZOI Universal Transmitter.

Although crystal controlled, the oscillator stage is a variable crystal oscillator (VXO). The crystal is a fundamental frequency crystal on 7.030MHz, the CW QRP calling frequency on the 40-metre band. Notice that an inductor (L1) and a variable capacitor (VC1) have been added in series with the crystal and ground. These inductive and capacitive elements allow some shifting of the crystal frequency; adding series capacitance shifts the frequency higher and adding series inductance shifts the frequency lower. The combination of L1 and VC1 does give a useful frequency variation for the oscillator. L1 is a 39μH axial choke. C1 is a section of a polyvaricon variable capacitor with a maximum capacitance of 60pF.

The VXO is link coupled, via T1, to a Class C power amplifier. T1 has a tuned primary winding with VT1. The tuned winding (35 turns of 30SWG wire on a T50-2 core) requires VT1, a pre-set capacitor, to resonate on the band. A five-turn link winding is wound over the centre of the tuned winding to couple the oscillator to the amplifier.

As with the original W7ZOI design, the oscillator is keyed and the amplifier is powered at all times. Some constructors prefer to key the amplifier stage via a transistor DC switch. The problem here is that the oscillator will run at all times and hence be heard in the accompanying receiver. There are other ways around this, but they would just add extra complexity to a very simple circuit.

The power amplifier is very simple. R5 provides a load for the signal from the link winding and drives the transistor into Class C. An RF load is produced by RFC1; another 39μH choke. C5 couples the RF output to a seven-element low-pass filter based on the W3NQN standard value capacitor (SVC) filter data. Capacitor C6 leads to the transmitter-receiver break-in circuit popularised by W7EL in his Optimised Transceiver design. This is a simple method of sharing the same antenna by a transmitter and receiver with- out damage to the receiver or significant loss of signal from the transmitter. However, it does not mute the receiver so the operator simply keys the transmitter and turns down the gain control of the receiver. It does have the advantage of being able to monitor the trans- mission with the receiver.

S1 is a "netting" switch that switches off the power amplifier. When the key is depressed the

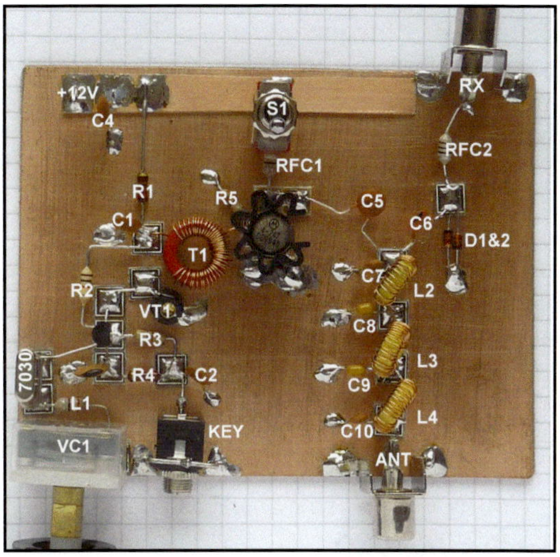

Layout of the Buildathon 40m transmitter

VXO runs but the board does not transmit. This facility allows the oscillator to be heard in the receiver so that the VXO signal can be adjusted to a desired frequency or aligned with a

desired station signal. Netting is standard practice when a separate transmitter and receiver are being used.

As stated earlier, the Buildathon transmitter uses Manhattan-style construction with small pads as connecting points for circuitry that is mounted over a conductive ground-plane (a piece of printed circuit board material). The pads are attached to the ground-plane using 'super glue', which sets in seconds to offer a very durable bond. Ground connections are soldered directly to the copper ground-plane. Each pad should be pre-tinned before adding the parts by heating the pad and melting solder to form a solder film over the pad.

Building the transmitter

Begin the transmitter by building and testing the VXO. This involves adding the parts as far as R5. The first task is to glue the 'bus bar' in place. This cop- per strip carries the 12-volt supply for the transmitter. The copper strip should be placed across the top left corner of the printed circuit board base. If the left side of the strip is against the left edge of the board, there will be enough space in the top right corner for the receiver (RX) phono connector.

The VXO tuning capacitor (VC1) is a polyvaricon type variable capacitor commonly available from electronic component traders. It has three connections, marked 'O' for oscillator, 'G' for ground and 'A' for antenna. The antenna (A) section has a maximum capacitance of 140pF and the oscillator section has a maximum capacitance of 60pF.

The VXO requires the use of the 'G' and 'O' connections. It is a good idea to snip off the 'A' tab. VC1 is mounted in the bottom left corner of the board. The 'G' tab is soldered directly to the ground-plane and the 'O' tab is connected to a pad, as shown in the photograph. It is helpful to make a blob of solder on the board before soldering the 'G' tab in place. VC1 is then bent backwards so that it is at right angles to the board.

The 7030kHz crystal is mounted in a holder made from a small section of integrated circuit header pins. Three pins are required with the centre pin cut off and the outer two pins providing the correct spacing to mount the wire ended crystal. The outer pins are soldered to a pair of adjacent pads. Cut the wires on the crystal to about 5mm. The crystal should push into place in the holder.

The photograph shows the layout of the VXO circuit. Please note that R5 is missing from this photograph and should be added between the output pad of T1 and ground.

The 2N3904 VXO transistor is mounted on three pads, the flat side of the transistor pointing towards R3.

T1 is wound on the T50-2 (red) core. The tuned winding has 35 turns of 28SWG wire (about 24in of wire is required). Each time the wire passes through the hole of the core counts as one turn. The turns should be laid side by side and occupy about three-quarters of the circumference of the core. The link winding is five turns of wire wound over

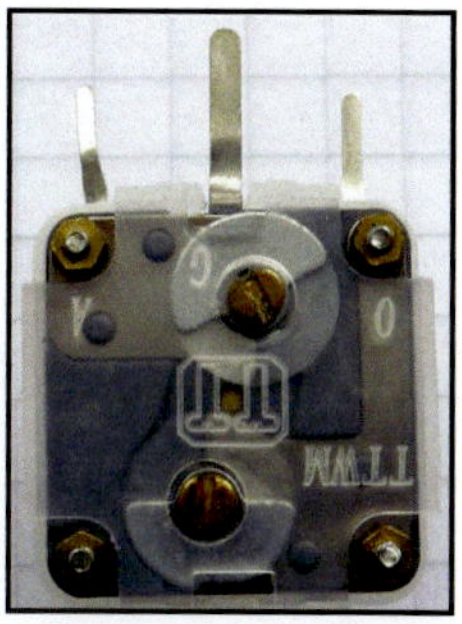
VXO tuning capacitor

Connections for the VXO variable capacitor

Construction of toroidal transformer T1

the centre of the main winding. Placement of the link winding is not too critical - just guess the centre position. When the windings have been made, trim the ends of the wires to about 1 cm. The wire is coated with enamel insulation. About 5mm of the enamel coating on the ends of the wire must be scraped clean to expose the bare copper and tinned with solder. The best method is to form a blob of solder on the soldering iron tip, apply the tip to the end of the wire and wait until the solder flows onto the copper wire. The tuned winding is attached to two pads and the link winding to one pad and ground.

Testing the VXO

Once the VXO stage has been built, it can be tested using the circuit shown in **Fig 22.2**. At the same time VT1 can be adjusted for maximum VXO output. This will require the use of a diode RF Probe. This is a simple test item to build with two 100nF capacitors and two 1N5711 diodes. Ideally the meter should be an analogue type (moving needle) as this shows the peak output voltage better.

The probe is connected as shown between the VXO Test Point and ground. 12V from the power sup- ply is applied to the top left-hand corner of the power bus bar. Positive (+) to the bus bar and negative (-) to the ground board (it helps to solder a small piece of wire to the bus bar to accept the clip lead from the power supply).

Plug in a Morse key with a 3.5mm jack plug. Depressing the key will switch on the VXO. Adjust VT1 for maximum output voltage. Note that the VXO may not oscillate until the output is peaked.

Builders may also listen to the VXO on a receiver and measure the frequency with a frequency counter if one is available. The most difficult part of the transmitter is now built.

Close-up of the VXO layout

Completing the transmitter

Have the picture of the whole transmitter to hand to show where the components are placed (the picture

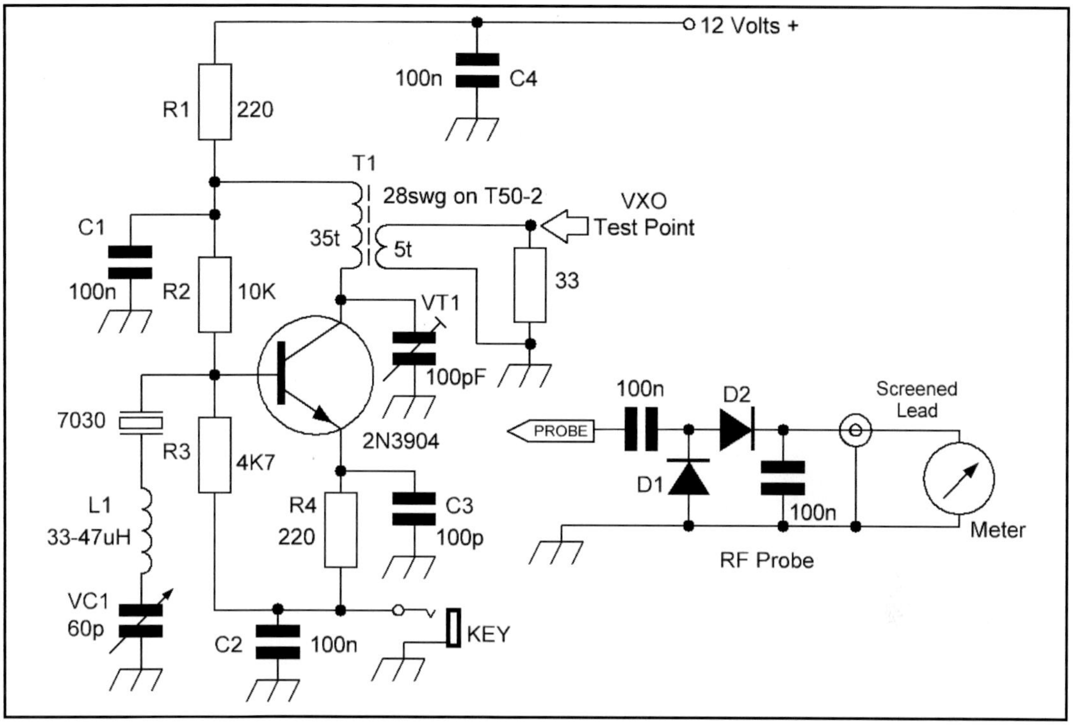

Fig 22.2: Testing arrangement for the VXO

labelled "Layout of the Buildathon 40m transmitter" earlier in this chapter). Add the amplifier transistor (the large one) RFC1 and C5. When mounting the transistor, it helps to bend a little of the leads to form a 'foot' to lay flush on the pads.

The picture (right) shows the detail around the transistor, S1 and RFC1. Two pads, connected together, are used for the collector lead of the transistor. This gives more room for the parts. The top of RFC1 is connected to a pad mounted against the power bus bar. This provides a pad for the "NET" switch SW1. Solder RFC1 in place before S1.

The switch has three tags. Cut off one of the outer tags and solder the switch into place as shown. The purpose of S1 is to switch off the supply to the output transistor. This allows only the VXO to be keyed to provide a signal for netting the transmitter.

The heatsink is applied to the output transistor by placing a screwdriver blade in the gap and twisting it to open the vanes. The heatsink should now fit over the transistor. The screwdriver is removed to allow the heatsink to clamp firmly on the sides of the transistor.

Wind the coils for the low-pass filter. These are wound on the yellow toroids and each will require about 15" (38cm) of 28SWG wire; 21 turns for L2 and L5, 24 turns for L3. The low-pass filter is mounted on five evenly spaced pads; the coils being at an angle to fit the spacing.

The transmitter output and the receiver input use phono sockets. The tabs of the sockets need bending before mounting on the board. The outer two tabs are bent outwards at 90 degrees and the smaller, centre, tab is bent inwards. The socket will now mount flush with the board as shown in the picture of the completed transmitter. Make good islands of solder on the board and tin the outer tabs before soldering the socket in place. When the transmitter board is completed, check the wiring of the final stages and filter.

The transmitter can now be tested on the test bench. Do not key the transmitter without

Fitting S1 and RFC1

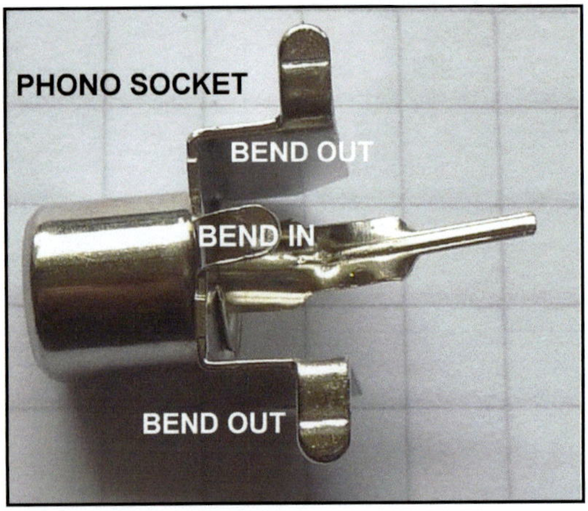

Preparing the antenna socket for placing on the board

it being connected to an antenna or a dummy load. An RF probe with a dummy load will be needed. Measure the output voltage - this is peak to peak voltage. Divide this voltage in half, multiply it by 0.707 to give the RMS (root mean square) voltage. The output in watts is the RMS voltage squared and divided by 50. The output should be in the order of one watt; quite a useful low power output on 40 metres. If the out- put is low try replacing R5 with a 39-ohm resistor to give more output.

The full components list is shown in the table below. Most of the parts are commonly available from electronic components traders. The 7030kHz crystal is available from the GQRP Club (www.gqrp.com) and some component traders. Ready-made pads are available from QRPme (www.qrpme.com).

TRANSMITTER PARTS LIST	
R1	220 Ω
R2	10k Ω
R3	4k7 Ω
R4	220 Ω
R5	33 Ω
C1	100nF (0.1 μF)
C2	100nF
C3	100pF
C4	100nF
C5	100nF
C6	47pF
C7	270pF
C8	680pF
C9	680pF
C10	270pF
VC1	60pF polyvaricon
VT1	100pF trimmer Murata 5mm black
T1	T50-2, 35t + 5t
L1	33 μH, axial choke
L2	T37-6, 21t
L3	T37-6, 24t
L4	T37-6, 21t
RFC1	33 μH, axial choke RFC2 10 μH, axial choke
Wire	28SWG
D1	1N5711
D2	1N5711
TR1	2N3904
TR2	2N3866
Key jack	3.5mm, phono
Antenna jack	phono
Rx antenna jack	Phono
Crystal	7.030MHz
Blank PCB	10cm x 8cm
12v bus strip	7.5cm x 5mm
Heatsink	
Knob	
Crystal socket, DIP strip	
Super glue	

The Sudden - a kit transmitter

At the time of writing a kit version of a similar transmitter is available from the G QRP Club (http://www.gqrp.com/sudden2.htm). The circuit is a VXO, followed by a buffer and the power amplifier is a 2N3866. The output is around two watts, the VXO and buffer are keyed by another transistor and there is muting on the receiver antenna feed. There are four toroidal coils to wind - a coupling transformer and the three inductors of the seven-element low pass filter.

The kit uses a new method of construction, based on the 'Manhattan' idea, but called Limerick. With this method, developed by Rex Harper, W1REX, (who lives in Limerick, Maine), the pads are etched on the board, and the interconnections between them are PCB tracks, hidden under the solder masking. No 'through hole' connections are used, although 'through hole' components are used.

The kit includes all components, hardware, and a case. All you need to supply is a soldering iron and solder, and a few simple hand tools. The iron should be of a reasonable size, as the case is made by soldering the parts together and a low power iron will not be man enough.

The front of the Sudden Transmitter, so called because it was designed as a companion to the Sudden receiver

The inside of the transmitter shows the method of construction. The parts are soldered to pads in the Manhattan construction style. The interconnections between the pads are hidden under black solder masking. All pads are marked showing where to connect the components

The Buildathon receiver (a simple 40m receiver)

The Buildathon Receiver is based upon the Sudden direct conversion receiver design. The Sudden receiver is built around two integrated circuit chips; the NE602 and the LM386. Both lend them- selves to simple receiver construction; although neither has particularly high performance in that application. Having said that, the Sudden is a remarkable receiver for what it contains. The original Sudden began life over 20 years ago in SPRAT and was then published in Practical Wireless followed by the 73 magazine in the USA. Since then several versions and modifications to the Sudden receiver have been written up; although very few of them were written by me. The original naming of the 'Sudden' had nothing to do with rapidity or hastiness but was taken from the name of the place where I lived and worked. The SPRAT version was available as a kit from Kanga Products. The kit proved very popular with beginners and several groups used it as a training project with young people. There was even a surface mount version; produced for a while by Blue Rose Electronics. In more recent times a version of the receiver was packaged in a tuna tin on a circular printed circuit board and called 'The Sudden

Storm' by W1REX at qrpme.com.

The circuit (**Fig 22.3**) is a simplified version of the Sudden receiver that uses a VXO (variable crystal oscillator) tuning around the QRP calling frequency (7.030MHz) on the 40-metre band. The Sudden has only two active components: U1, an NE612 (or SA602) chip which contains a mixer and an oscillator and U2, an LM386 audio amplifier chip. The signal at the antenna input goes directly to an attenuator; a 10kΩ linear potentiometer (R1) that serves as the receiver gain control. R1 is the only gain control on the receiver; the audio amplifier runs at full gain. In a simple receiver the best place for a gain control is at the antenna input. The required signals are selected by an input tuned circuit T1/C1 and C2. The link winding of T1 receives the signals from the gain control. The output from the tuned circuit goes to the input of the mixer at pins 1 and 2 of U1.

U1 contains an internal oscillator circuit that is accessed via pins 6 and 7. This is a version of the Colpitts oscillator; the frequency of which is determined by a 7.030MHz crystal. The frequency of the crystal can be shifted by L1, an axial choke, D1 and D2. D1 and D2 are 1N4005 silicon diodes but here they function as varactor diodes. A varactor (or Varicap) diode is a diode which has variable capacitance controlled by the voltage on its terminals. The voltage feeding D1 and D2 comes from a potentiometer (R4) via R3. R4 must be a linear track potentiometer. Adjustment of R4 alters the voltage and hence the capacitance across D1 and D2 shifting the frequency of the crystal, allowing a small tuning range around 7.030MHz.

The mixer section of U1 mixes the signals from the input tuned circuit and the signal from the oscillator and the resulting signals appear at pins 4 and 5.

An audio (sound) output will appear at the upper and lower side- bands of the oscillator frequency. These are the signals we require. The audio signals are fed to U2 via C12 and C13; this uses the balanced input to the LM386. The capacitors C9, 10 and 11 form a simple filter to give some audio tailoring. C8 between pins 1 and 8 of U2 (the gain control pins) gives

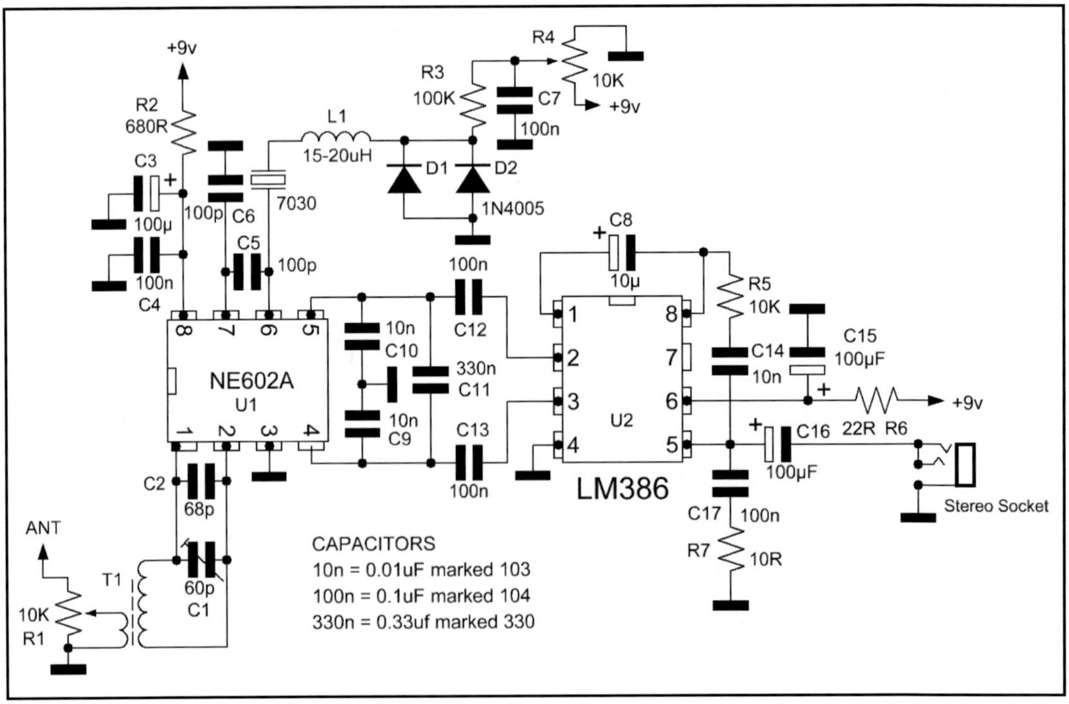

Fig 22.3: The complete circuit diagram of the Buildathon version of the Sudden receiver

A Sudden receiver built at a 'buildathon' in the UK

the maximum voltage gain of about 45dB. The output appears at pin 5 and is fed to the output jack via C16. C17 and R7 which form a filter (called a zobel network) to aid amplifier stability.

The output of the LM386 is about 350 milliwatts. This will drive a small loudspeaker although it works better with 'Walkman-type' headphones or earbuds.

Building the receiver

The receiver is best built 'in reverse' beginning with the audio amplifier section. This section can be tested before the rest of the receiver parts are added. The audio amplifier section is those parts around U2 to the right of C12 and C13 on the circuit diagram.

U2 will require an 8 pin DIL (dual-in-line) carrier for mounting the chip. These are easy to make although commercial ones are available. Attach the carrier to the board with super glue. Tin (cover with solder) the pads on the carrier. The integrated circuit (IC) chip can be soldered directly to the carrier but adding an IC holder is a good idea as it allows the chip to be plugged into the carrier.

It is important to get the correct orientation of the pins on the LM386; pin 1 is the top left pin in the photograph (see also the diagram below). The electrolytic capacitors (C8, C15, C16) must also be mounted correctly. The negative side, indicated by the black, filled in, side on the circuit symbol usually has a black bar on the side of the capacitor case.

The audio amplifier requires a 9V (PP3) battery for testing so the battery leads need to be connect- ed. To test the amplifier, plug a small loudspeaker or pair of Walkman style headphones/earbuds into the 3.5mm jack socket and connect the 9V battery. There should be a slight rushing sound. Touching pin 2 or 3 should produce some hum. If all is well, the rest of the receiver can be built.

The next section to be built is the connection between U1 and U2. Place the IC carrier for U1 about 3cm away from U2. Then add C9 to C13. C9 and C10 can be placed on either side of C11 as shown in the photograph. The outer leads of C9 and C10 are connected to ground.

Next add the parts associated with the variable crystal oscillator. The layout is a little tight around pins 6 and 7 of U1. C6 is best soldered to the left-hand lead of C5 rather than pin 7. Pads are required for the crystal and R4. The photograph shows an edgewise potentiometer used for R4. If a conventional potentiometer is used wires may have to be added to connect to R4. Diodes D1 and D2 are mounted vertically, side by side, with the bottom ends connected to the ground. The white bar on one end of the diodes indicates the cathode, the upper- most end of the diode that connects to L1 and R3.

The final section to be added is the input and tuning section around T1 and C1. T1 is wound on a T50-6 toroi-

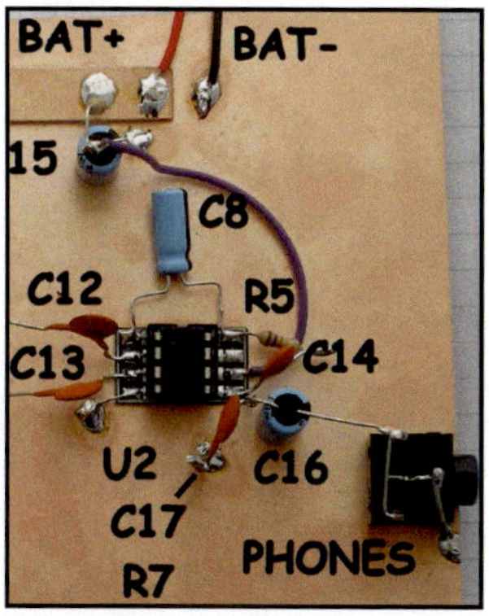

The audio amplifier

dal core. This looks a little like a red coated Polo (aka life saver) mint. The main winding is 33 turns of 30SWG enamelled copper wire. Each time the wire passes through the core counts as one turn. The input winding of T1 is 4 turns of the same wire wound over the centre of the main winding. The exact placement of this winding is not critical. Leave about 1.5cm of wire at each end of the windings and scrape off the enamel to reveal the bare copper. Tin the wire and solder it to the pads as shown.

The VXO tuning capacitor (VC1) is a polyvaricon type variable capacitor commonly available from electronic component traders. It has three connections, marked 'O' for oscillator, 'G' for ground and 'A' for antenna. The antenna (A) section has a maximum capacitance of 140pF and the oscillator section has a maximum capacitance of 60pF. C1 requires the ground (G) and oscillator (O) connections soldered to the pads connected to the main winding of T1. R1 and the antenna socket are placed on the edge of the board. Again, the illustrated board shows an edgewise potentiometer used for R1.

The components shown in the parts list are commonly available from electronic components traders. The 7030kHz crystal is available from the GQRP Club (www.gqrp.com) and other traders. Readymade pads and IC carriers are available from QRPme (www.qrpme.com).

The receiver can now be tested. Because T1 has a link winding, the receiver is designed for use with a low impedance antenna input; a dipole or other antenna via an antenna tuner. Having said that, I recall testing some of these receivers using a piece of wire about 3 meters long connected directly to the input . . . and we heard signals. The signals are peaked by adjusting C1 with R1 acting as a volume control. Although the receiver is simple and the tuning range is limited, it is a surprisingly viable receiver for listening to amateur stations around the international QRP calling frequency on 40 metres. There are, however, several improvements that could be made if you wish, and these are detailed below.

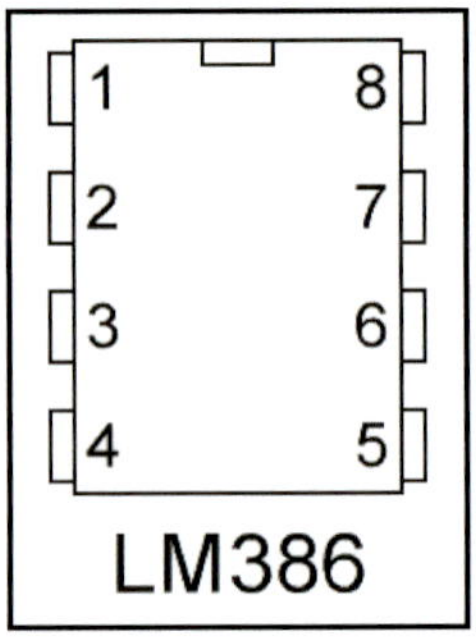

Pin numbers of the audio amplifier chip

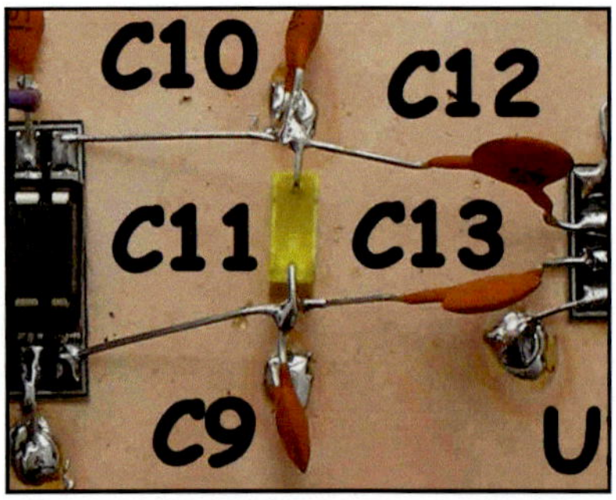

The connections between U1 and U2

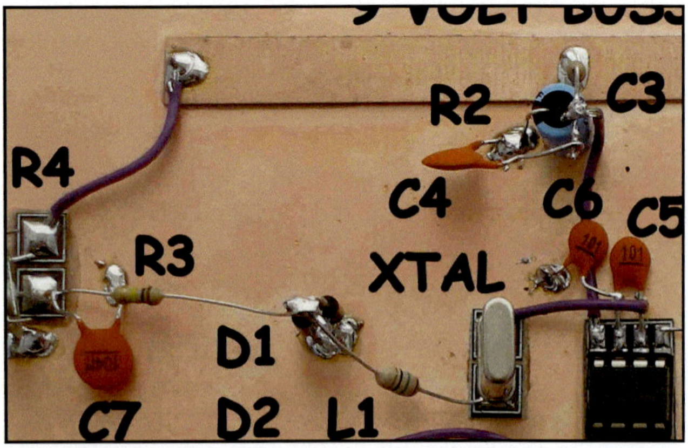

The VXO section

Adding a variable frequency oscillator (VFO)

Replacing the variable crystal oscillator (VXO) with a VFO (**Fig 22.4**) will enable the whole 40 metre band to be tuned. In this version, the frequency of the oscillator is determined by L2 and its associated capacitors, and adjusted by the 60pF tuning capacitor. The oscillator is tuned across the required amateur band. The oscillator coil, L2, is a 2u6LC component from Spectrum Communications. It is designed as a 10.7MHz IF transformer and includes a built-in capacitor which is a ceramic cylinder with metal end caps located in the base of the coil. This capacitor must be removed before L2 is used. It can easily be removed by inserting a sharp

The input and tuning circuitry

screwdriver and breaking the ceramic centre. Insert the screw driver blade on the side away from the centre core of the coil former and gently twist or prise the blade until the ceramic tube breaks away. Note - a 2u6L coil is now available which does not have an internal capacitor.

Ideally a frequency counter and signal generator may be used to set up the VFO. For those without these items of test equipment, it is quite possible to set it up 'by ear'.

Using a frequency counter to set up the receiver oscillator:

L2, the oscillator coil, has an unused link winding. This may be used to feed a frequency counter. Connect one side of the link winding (either side) to ground and use the other end to connect to the counter. Some counters may detune the frequency a little so putting a low value capacitor (47 to 100pF) between the counter and the link winding is advised. Set the tuning capacitor control fully anti-clockwise and adjust the core to read a frequency just below 7.0MHz. Be warned! Rough use of the adjustable core can break it. Ideally use a proper plastic 'trimming tool' in the slot on the core. If a trimming tool is not available, use a small screwdriver . . . with care.

Using a signal generator to set up the receiver oscillator:

Set the signal generator on 7.0MHz and feed the signal into the antenna input. Begin with a low output from the signal generator and only increase the output if it cannot be detected by the receiver. Set the tuning capacitor control almost fully counter-clockwise and gently rotate the core of L2 until the generator signal is heard.

BUILDATHON RECEIVER PARTS LIST	
C1	60pF variable capacitor
C2	68pF, marked 68
C3, C15, C16	100μF electrolytic
C4, C7, C12, C13, C17	100nF (0.1uF), marked 104
C5, C6,	100pf, marked 101
C8	10μF electrolytic
C9, C10, C14	10nF (0.01uF), marked 103
C11	330nF, marked 330
D1, D2	1N4005 diode
R1, R4	10k Ω linear taper potentiometer
R2	680 Ω, blue-grey-brown
R3	100k Ω, brown-black-yellow
R5	10k Ω, brown-black-orange
R6	22 Ω, red-red-black
R7	10 Ω, brown-black-black
L1	15 or 20μH axial choke (choice)
T1	T50-2 core - 4t pri; 33t sec
U1	NE602 (or LM612)
U2	LM386
Crystal	7030kHz
Blank PCB	15cm x 10cm
12v bus strip,	10cm x 5mm
3 pin SIP xtal socket	
30SWG enamelled copper wire	
3.5mm headphone connector	
9-volt battery snap lead	
9-volt battery (PP3)	
2 8 pin DIL sockets and 8 pin IC carrier	
10 solder pads	
2.25" pcb strips	
1 tube Superglue	

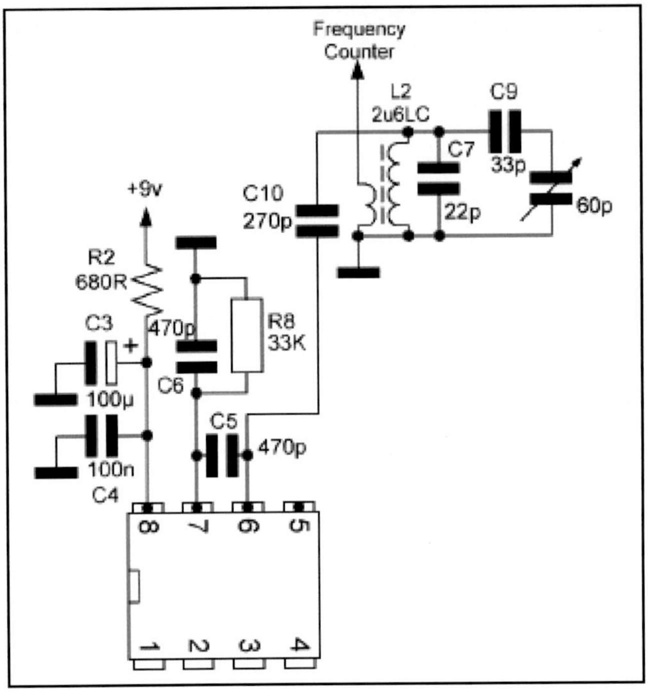

Fig 22.4: The VFO circuit

Removing the capacitor from the coil unit

Setting the oscillator frequency by ear:

Connect an antenna to the receiver input (a good antenna is helpful) and set the tuning capacitor nearly anti-clockwise. Adjust the core of L2 until CW (Morse) signals on the 7MHz band can be heard. The strength of the signals will depend upon the time of day. It will require some experimentation to find the low end of the band. Try to find the lowest frequency CW signal.

Adding better input tuning

In this improvement (**Fig 22.5**) the required signals are selected by a two-stage band-pass filter based on two Spectrum Communications 5u3L coils. There are two tuned circuits L1/C2 and L2/C3 loosely coupled by C2. Note that the tuned windings of L1 and L2 have a centre tap connection that is not used. Only the two outer connections are used. The link winding of L1 receives the signals from the gain control but the link winding of L2 is unused, the output being taken from across the tuned winding. The output from the band-pass filter goes to the input of the mixer at pins 1 and 2 of IC1.

To set up the input filter, set the two input tuned circuits L1/C1 and L2/C3 to maximum

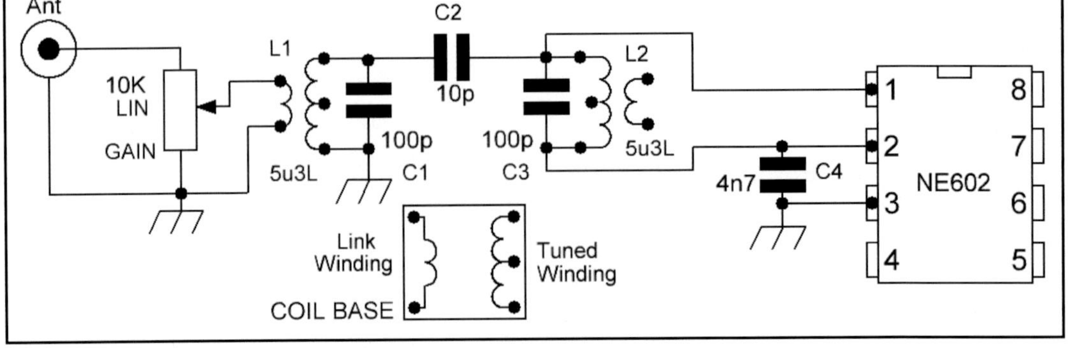

Fig 22.5: The modified input circuit

output using the cores in L1 and L2. Use a trimming tool or work very carefully with a small slotted screw- driver. L1 and L2 can be peaked for maximum signal strength using a signal generator or amateur band signals. Even if using a signal generator, it is an advantage to do a final adjustment listening to signals on the band; as these are what we will be using the receiver for. Use the best available antenna for these adjustments. Adjust L1 first for a peak in the signals and then adjust L2. Repeat this several times until the best results are obtained. Take care not to screw either core right through its coil!

The position of the core will vary with individual receivers because of the tolerance of C1 and C3 and the nominal inductance of individual coils. Usually the input filter will peak with the core set at about two complete turns downwards from the position where the core is flush with the top of the coil former. Remember that the filter is based on a nominal input impedance of 50 ohms, so a 50-ohm impedance antenna or an antenna tuning unit will obtain the best results.

A kit version

At the time of writing a kit version of the receiver, with bandpass tuning and VFO coverage of the full band is available from the GQRP Club (www.gqrp.com/sudden.htm).

This club version of the Sudden is a re-think of the original G3RJV design, with a VFO to cover the full band, and minor improvements to the circuitry. The kit uses the 'Limerick' method of construction,

as used in the transmitter kit. The pads are actually etched on the board (like surface mount), and the interconnections between them are PCB tracks, hidden under the solder masking. No 'through hole' connections are made, although 'through hole' components are used. The kit includes all components, the complete case, a knob and even a PP3 battery. All you need to supply is a soldering iron and solder, and a few simple hand tools. The iron should be of a reasonable size, as the case is made by soldering the parts together, and a low power iron will not be man enough.

The completed receiver needs a reasonable antenna, and a pair of Walkman style earphones, although the audio stage will drive a small loudspeaker in a quiet room. The original Sudden kit was for 40m, but these club kits are available in 160m, 80m, 40m, 30m, or 20m versions.

A Z-Match antenna tuning unit

The Z-Match ATU described here (**Fig 22.6**) is a version of the SPRAT circuit by G3WQW that has been copied many times and formed the basis of several kits. It uses one coil and two polyvaricon variable capacitors. The table overleaf shows why this ATU is popular with radio amateurs.

Using a toroid inductor and small variable

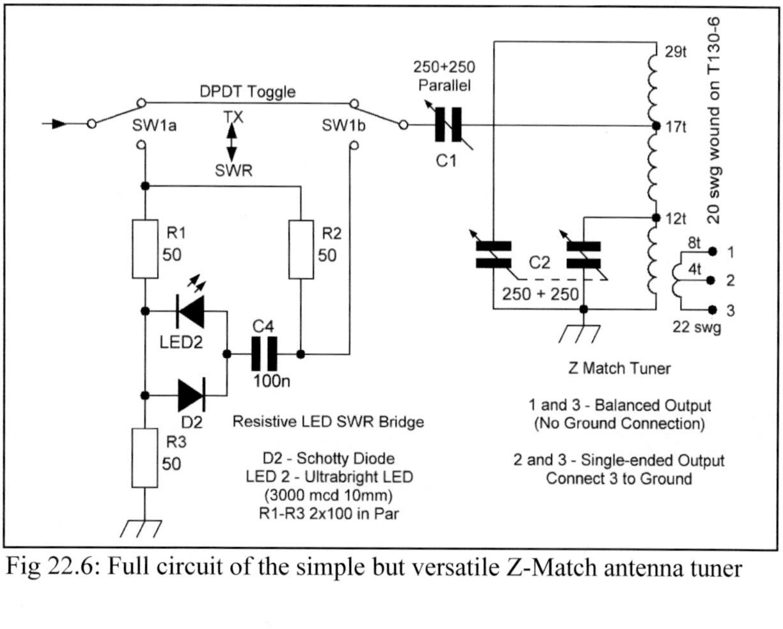

Fig 22.6: Full circuit of the simple but versatile Z-Match antenna tuner

The GQRP Club kit version of the Sudden receiver includes a VFO and fits into this neat case. It is available in versions for 80m, 40m, 30m or 20m

capacitors, this Z-match can be built into a very compact package. With the components used, it is only suitable for QRP power levels.

The first stage is to attached the PCB front panel to the ground plane (base plate). It helps to use a wooden block to align the front with the base. It is a good idea to secure the front panel with a large blob of solder at each end. Another blob of solder is then used to secure the centre of the join. To complete the join run a line of solder along the whole joint. This does require a hot iron with a large bit.

The signal from the transmitter passes through a Resistive SWR Bridge. Because the bridge will reduce the output from the transmitter a double pole switch (SW1a & SW1b) is used to short out the bridge when the transmitter is in use. The best match between transmitter and antenna occurs when the bridge indicates a null seen as the dimming of the LED.

The signal is coupled into the Z-Match via C1; a variable but high value of capacitance. The Z- Match is a parallel tuned circuit with link coupling to the antenna. The two variable capacitors with the tapped tuning coil offers a wide tuning range. It should be possible to use the tuner on all HF bands from 80 metres to 10 metres to match balanced or single ended antennas. The whole link winding is isolated from ground (using connections 1 and 3) and hence will feed a balanced antenna.

The Z-Match tuner for QRP can be fitted into a compact space

Joining connection '3' to ground and using connection '2' will tune a single ended antenna such as a long wire, against ground. To use the Z-Match with a single ended or unbalanced antenna, connection '3' must be connected to ground. This will require adding a link wire between '3' and ground. Some antenna configurations require a higher capacitance value of C1 to obtain a good match and some examples of this tuner offer extra capacitance that can be added to C1. How to include this option is discussed in the notes at the end of this documentation.

ADVANTAGES AND DISADVANTAGES OF THE Z-MATCH

Advantages

- Matches balanced loads without the use of lossy baluns.
- Being a parallel resonant circuit, the Z-match can provide some band-pass filtering for your receiver and harmonic attenuation for your transmitter.
- A well-designed Z-match tuner has a high Q and is more efficient (less lossy) than other types of tuners.
- The fixed inductor simplifies construction (no switches or rollers needed).

Disadvantages

- Tuning is usually very narrow so it can sometimes be a bit touchy to tune up

Mounting the variable capacitors

Fasten the two variable capacitors to the front panel. The best method is to put a nut onto each of the small bolts (on the inside) Screw the nut almost to the front panel. Introduce the threaded holes of the capacitor to the bolts and lightly screw up each end. When they are aligned - tighten up the bolts. The solder tags should point downwards.

The SWR indicator pads and phono sockets

The parts for the SWR Indicator are mounted 'Manhattan-style' on square pads. The layout is shown in the photograph. There are seven pads, six of these being mounted in pairs with a wire bridge. The pads are held in place with superglue. Place one spot of glue on the base of the pad and apply it to the ground plane and hold firmly in place for about 20 seconds. The photograph shows how the parts will be placed on the pads. This is best done later when the switches are in place.

The input to the tuner is via a 'phono' socket. This can be changed later, if required, to match the usual plugs and sockets of your station. Two more phono sockets are mounted at the other end of the panel for the balanced output. These sockets can accept 4mm wander (banana) plugs. Alternatively, phono plugs may be used with just the centre pin soldered to two croc clips to grip balanced antenna feeder wires. The phono sockets have to be bent as shown so that they may be soldered on the ground plane.

Winding the coil - probably the most difficult bit

Fig 22.7 shows the completed transformer, the designations in brackets are a guide when winding the coil. The main coil has 29 turns of 20SWG wire, tapped at 12 and 17 turns. This requires about 50 inches (127cm) of wire. Each pass of the wire through the hole of the core counts as one turn. Leave about an inch (2.5cm) of spare wire at the beginning of the coil.

Keep the turns close to the core with a tight winding. After 12 turns, pull out a loop a wire (about an inch, 2.5cm long) and lightly twist it to hold the position. Continue to the 17th turn and make another loop. Complete the 29 turns, leaving about an inch (2.5cm) of wire at the end.

Snip the loop in the centre to form two

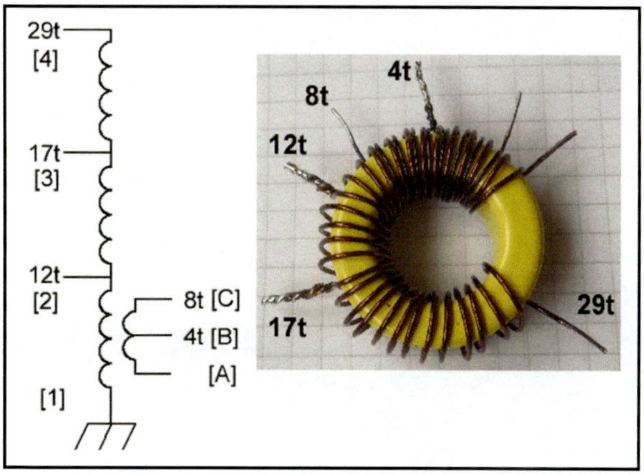

Fig 22.7: Coil schematic and construction details

wires - the position of the wire is maintained by one twist. Scrape the enamelled coating off the two short wires until bright copper is exposed. Tightly twist the two wires together and tin the wires with a generous coating of solder. Tin the beginning and ending of the winding after removing the enamel coating. Spread the winding to occupy about three-quarters of the core.

The link winding has 8 turns of 22SWG wire (about 15" / 38cm)) with a centre tap. This winding should roughly occupy the centre of the first 12 turns on the main winding so begin the winding two turns in from the beginning of the main winding. Interleave these windings with the turns on the main winding. (wind into the gaps). Wind the coil and make the tapping connections in the same way as was done for the main winding.

The coil is mounted on seven standoffs arranged around the circumference of the former to the left side (viewed from back) of the board. Place the coil in position on the ground plane and using a pencil or small felt tip pen, mark out the placements for the standoffs.

The standoffs are applied to the ground plane with superglue. Before applying the glue to the stand- offs, roughen the bottom of each standoff with sand paper. Pads may be used in place of standoffs.

The mounted coil is shown in the picture. Take care when soldering the coil leads to the standoffs as the plastic will melt if too much heat is applied. It doesn't really matter if the coil is not a good electrical connection to the standoff pillar because the connecting wires to be added can be soldered directly to the tinned coil leads.

Don't forget to solder connection [1] to the ground plane (I did!). Another easy ground connection to forget is the ground pin of C2. The photograph shows the left-hand variable

The Z-Match is built using Manhattan-style pads

Fitting the components onto the pads

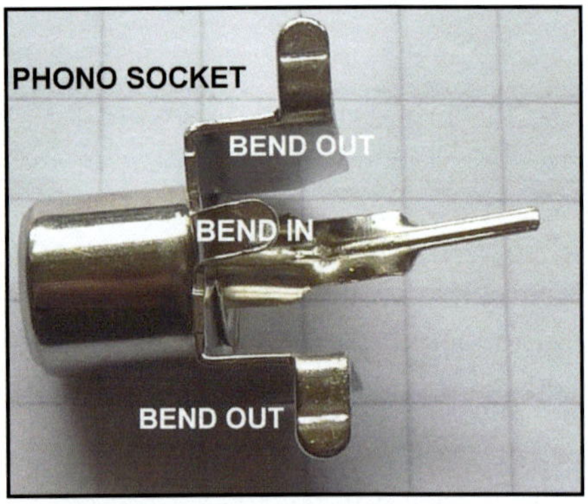

4mm wander (aka banana) plugs are compatible with the phono sockets

Modifying the phono sockets before fitting

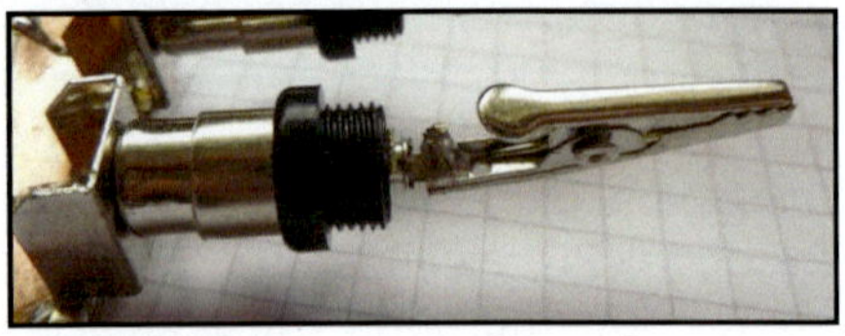

Alternatively a crocodile clip can be soldered onto a phono plug

capacitor viewed from the back of the board. The large centre tag is soldered to the back of the front panel. You may find it easier to use a piece of wire to make the connection as the tag is rather springy.

Wiring the SWR indicator

Add the two switches (S1 and S2) and the LED to the front panel. The LED is a push fit into the panel. S1 (the smaller switch) goes to the left of SW1a & b.

D2 must be connected the right way around with the black bar to the right. R1/2/3 are 50- ohms - made from two 100-ohm resistors in parallel. R1 and R3 are bent as shown for vertical mounting.

The longer (+) lead on the LED goes to the junction of C4 and D2. The shorter lead goes to the other side of D2.

The picture shows the wiring of SW1. The top two connections have a link wire to short out the SWR Bridge when transmitting. The middle left goes to the phono input socket. The middle right goes to the outer two connections of C1 (the two sections of the capacitor are in parallel). The bottom left goes to the junction of R1 and R2 and the bottom right goes to the junction of C4 and R2.

Wiring the rest of the Z-Match

The rest of the wiring necessary to complete the ATU is shown in the photograph. The wiring of the variable capacitors is shown below.

Some types of antenna appear to need more capacitance for C1.

I did not need this but the extra switch (marked S2 in the diagram; S1 on front panel) can be used to switch in an extra 470pF in

The coil is mounted on seven standoff insulators

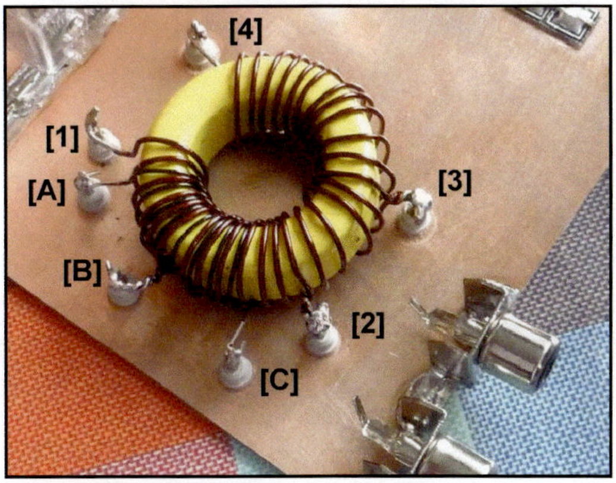

The completed coil in place.

Connections that are easy to forget.
[1]of the coil to the ground plane

[2] centre tag of C2 to the back of the front panel.

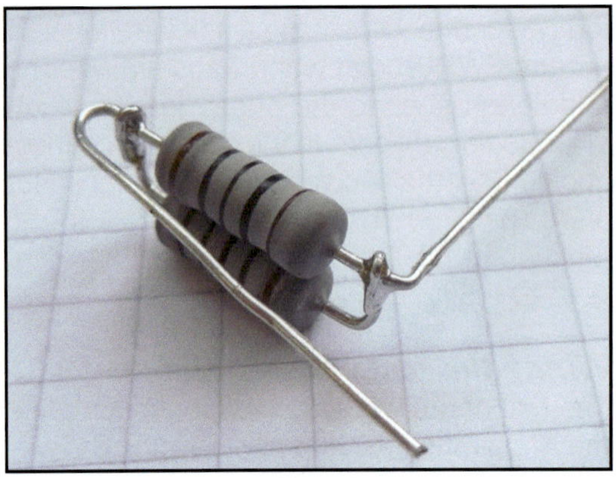

Making one 50-ohm resistor from two 100-ohm components

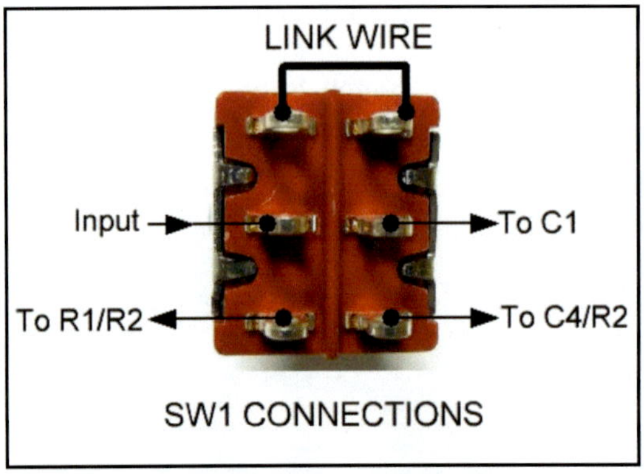

The wiring connections of switch SW1

parallel with C1 (**Fig 22.8**). If the switch is not needed for this application it could be used to switch [A] to ground when using the Z- Match with a single ended output.

Using the Z Match ATU

The Z Match is an excellent tuner for most types of antenna. The downside is that it can be tricky to use. Two useful tips are:

Keep a record of the readings of C1 and C2 for the best tuning settings for each band.

If possible do the initial set up using an in-line SWR meter as well as the built-in LED

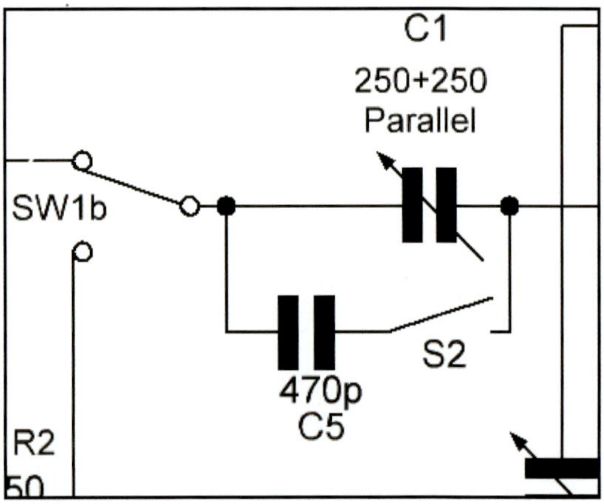

Fig 22.8: Increasing the value of C1

Placement of the SWR Indicator parts

indicator. The LED SWR Bridge only offers a simple indication of the state of tune.

Simple setting up procedure

- Connect the transceiver and antenna to the appropriate sockets of the ATU
- Set C1 at the centre of its range
- Listen for signals on the desired band
- Adjust C2 for maximum signal strength
- Apply a transmitted signal – no more than five watts
- Using BOTH hands (on C1 and C2) adjust C1 and C2 to obtain a null in the LED output
- If a null point cannot be found initially, try combinations of C1 and C2 until suitable settings can be achieved
- On some bands it may not be possible to completely extinguish the light of the LED. In this case aim for the lowest possible glow.

Z-MATCH ATU - LIST OF PARTS	
PCB Panel 10cm x 8cm	x 2
250+250 Polyvaricon Capacitor	x 2
T130-6 core	x 1
Wire 20SWG	50 inches
Wire 22SWG	15 inches
100-ohm 1w resistors	x 6
Schottky Diode	x 1
Ultrabright LED 10mm	x 1
100n capacitors	x 1
470p cap [optional]	x 1
Double pole change-over toggle	x 1
Single pole toggle	x 1
Insulated stand-offs	x 6
Phono sockets	x 3
Large knobs	x 2
MeSQUARES	x 6
Connecting wire	

Front panel of the Z-Match Tuner.

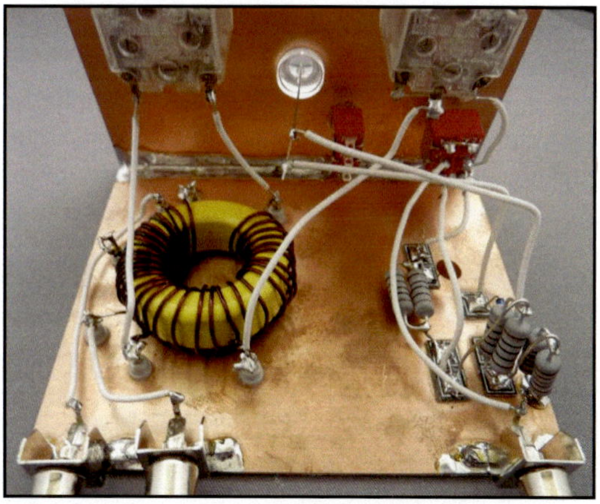
Wiring necessary to complete the ATU

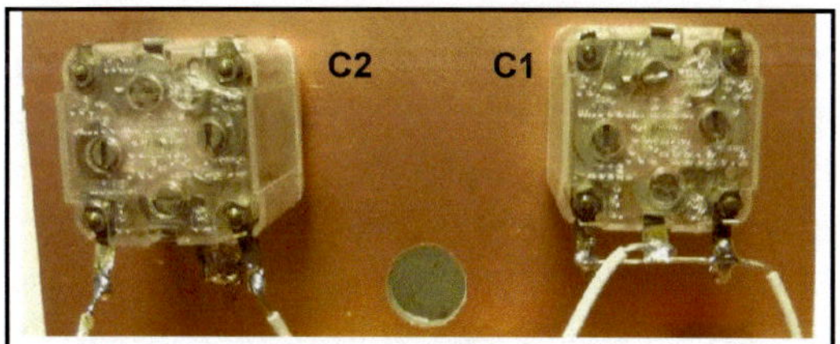

Coil [2]
Coil [4]
Coil [3]
SW1b Centre

The wiring of the variable capacitors

Chapter 23

Simple circuit building blocks for QRP

What this section tries to do is to show some useful simple circuits which can be built, either for use on their own or as part of a transmitter or receiver.

Emitter follower – a common buffer circuit.

The input impedance (usually taken as an input resistance) is approximately 0.5 R1, and the buffer's actual output resistance, Ro, for small signals, is taken as 25/Ie, where Ie is the emitter current in R3 in mA.

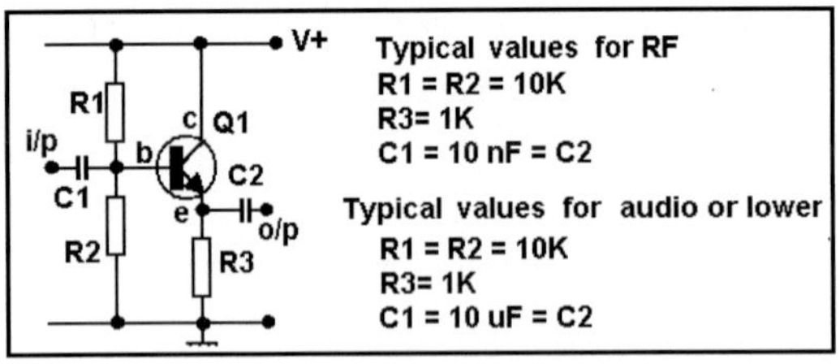

The N- channel JFET (NJFET) buffer

This is useful where loading of a previous stage has to be low. In this case, the circuit, also known as a "source follower", appears simpler, but it is a little harder to calculate the values to define the required bias, or the output resistance or impedance.

In this circuit, the formula relating gate-source voltage to drain current, that is, Vgs to Id, is given approximately, as below:

$$Id \approx Io \{1 - Vgs/Vp\}^2 \text{ -------}$$

where Io is the drain current for a given value of Vgs. Vp, the "pinchoff" voltage, is the value of Vgs for which Id is zero, which can be determined by measurement of the drain current when the

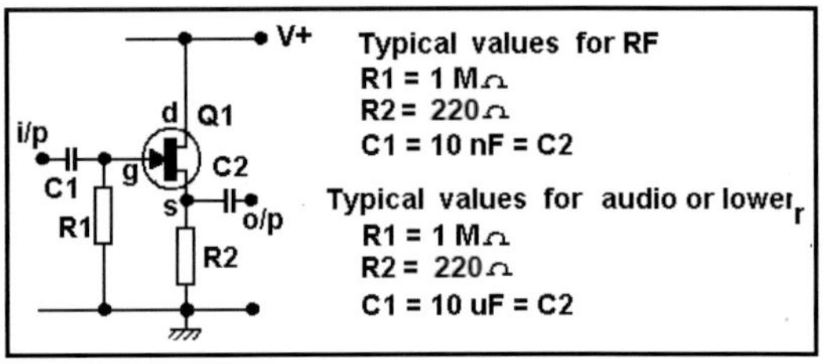

Vgs is set to zero, i.e. when R2 is zero. It does have the great merit of high input impedance, mainly limited by the resistance between gate and ground, which can be 100K and up to 10MΩ or more, and is often the basis for oscilloscope high impedance probes. The output resistance is higher than for the BJT circuit, being just a little less than R2. What has not been discussed is capacitance which can reduce the high frequency performance because of the high resistances involved at the input gate.

Common emitter amplifier 1

The circuit shown, with suitable values chosen for the supply voltage, the bias- and other resistors, is designed to have a good dynamic range for the signals appearing at the output. Typical NPN small signal transistors are used, and

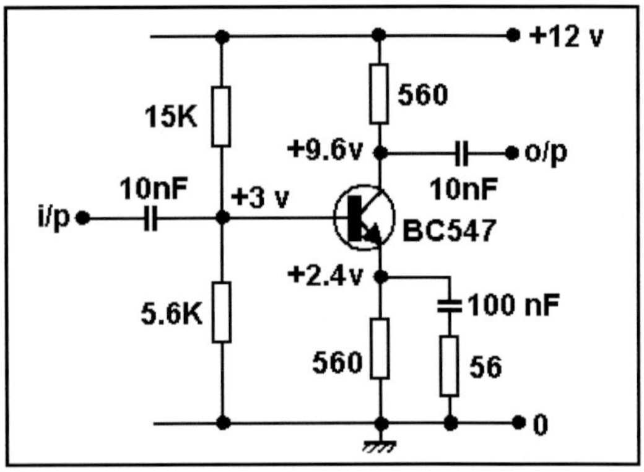

BC547 types can equally be replaced by such as 2N3904 ones for NPN, and to replace the complement to the BC547 types, being BC557's, then 2N3906 types can be used. Without the 56Ω resistor and the 100nF capacitor, the voltage gain from input to output would be -1, as the amplifier is inverting. However, above the "3 dB" frequency of the 100 nF capacitor and the 56Ω resistor (f3db ≈ 1/(6.3xCR), the voltage gain tends to -10. The base emitter voltage of the transistor is assumed to be 0.6 volts in this and other example circuits using NPN BJTs.

Common emitter amplifier 2

The minimum design of an inverting amplifier. In this case, the most easily constructed circuit can give very high gain, but the performance depends on the individual transistor's characteristics and the use of small input signals, as the amplifier can easily overload and distort the output.

The feedback between the collector of the BC547 transistor and its base ensures that the DC conditions are set up to keep the transistor biased correctly, with a base-emitter voltage of about 0.6 votls. The voltage gain depends on the resistances and also the individual transistor's performance in terms of current gain, which can vary according to manufacturing tolerances. Another point is that the input impedance is lower than in circuit 4 due to the feedback through the 330K resistance to the base.

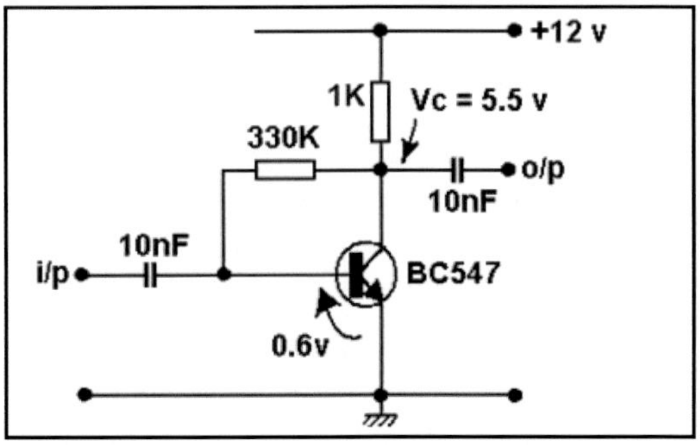

Stabilised voltage gain version with buffer.

In this version, the negative feedback is from an emitter follower buffer, which gives the desirable characteristic of a low output impedance (≈25/Ie for the second transistor) and an input impedance of around 10K. The point "X" is around 0.6 volts, and the point "Y" at 1.2 volts approximately. The voltage gain is -33 (i.e. inverting, given by the ratio of 330K/10K) and the circuit is stable, relatively independent of the transistor characteristics.

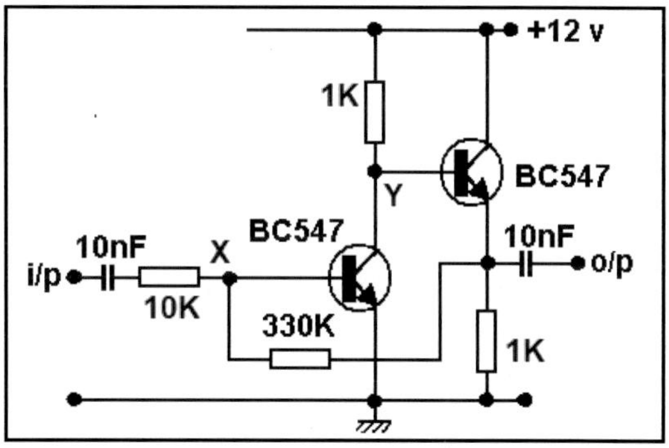

Common base amplifier.

This configuration has the desirable property of an input impedance/resistance that can be close to 50Ω, and with good gain-frequency characteristics. The circuit can therefore be used as an untuned front-end preamplifier in radio receivers. It is adapted from that of section 3 above, except the input is fed to the emitter and the base is grounded to signals by the 100nF capacitor as shown. The voltage gain can be quite high, and the amplifier is non -inverting. The input resistance is given by ≈25/Ie, and is thus 25/4 = 6Ω using the values in the circuit above. Because the convention is to have inputs on the left of the circuit, and outputs on the right, the common base circuit is often drawn as below:

For completeness, a power supply decoupling capacitor has been added, as shown.

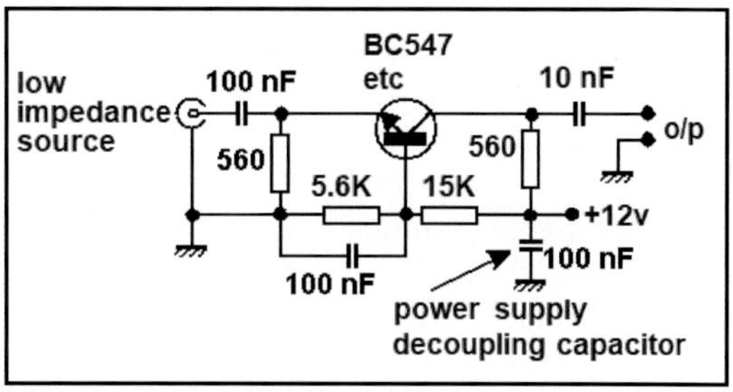

Tuned amplifiers – common emitter version for the HF bands.

Tuning in RF amplifiers is very important in order to set up the selectivity, interference protection, and noise minimisation. It is proposed only to give a couple of examples, in order to demonstrate the principle of tuning at the output. Amplifiers can be tuned at the input also, but this is not generally not both at the same time, as this can lead to unwanted oscillation.

In this example, a few extra features have been added. Firstly, there is the combination of L and C which forms a tuned circuit, the resonant frequency in Hz being given by:

$$f = \frac{1}{6.3x\sqrt{(L.C)}}$$

where L is the inductance in Henrys and C is the capacitance in Farads.

Thus, a capacitance of 100pf with resonate with an inductance of 1μH at a frequency of 15.9MHz. At the output of the amplifier, as shown, there should be a high input impedance circuit to follow.

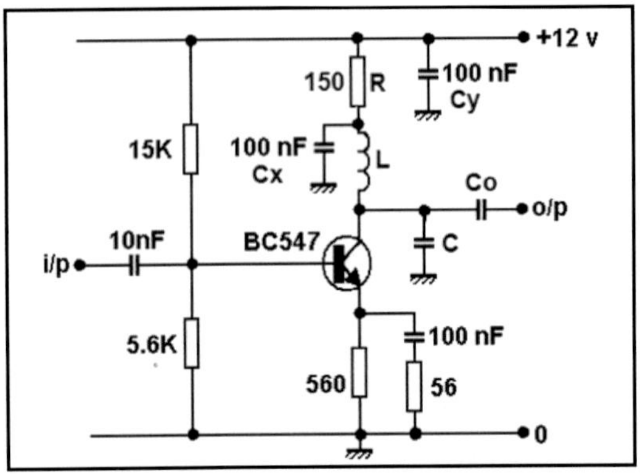

Tuned amplifiers – common base version suitable for higher frequencies in the HF bands.

The tuning concepts included here are the same as in the tuned common emitter amplifier, and so will not be detailed. Note again the supply decoupling circuit now includes a resistor and two capacitors, and that the BC547 has been replaced by a better, high frequency transistor which would allow more gain to be available at the higher frequencies of interest.

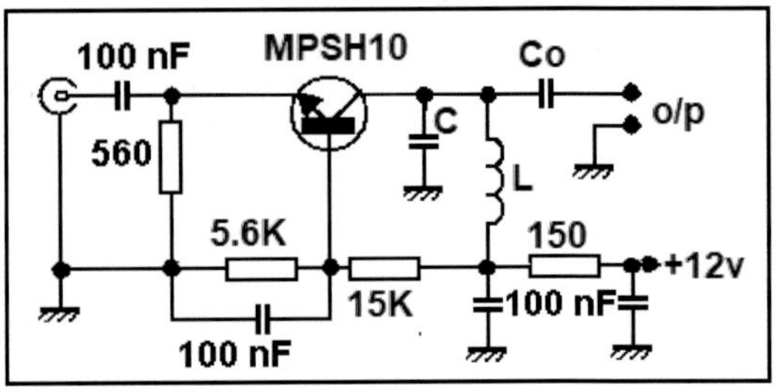

Simple oscillator circuit

There are several basic types, each with its own properties and relative advantages. A circuit can oscillate if it has, in effect, feedback between output and input of the right amplitude and phase at a particular frequency. Probably the most common one is shown below:

The output frequency depends on L, C, C1 and C2, neglecting initially the effect of the bias resistors of value 15K. Note that a crystal oscillator is implemented by replacing L by a crystal (or a ceramic resonator), which gives an output which is stable in frequency to a much higher degree. Variation of C allows frequency variation of the above circuit, which is smaller in extent if the circuit is a crystal- or ceramic-based oscillator. Increasing the 1.5K enitter resistor improves the waveform, but also increases the output impedance.

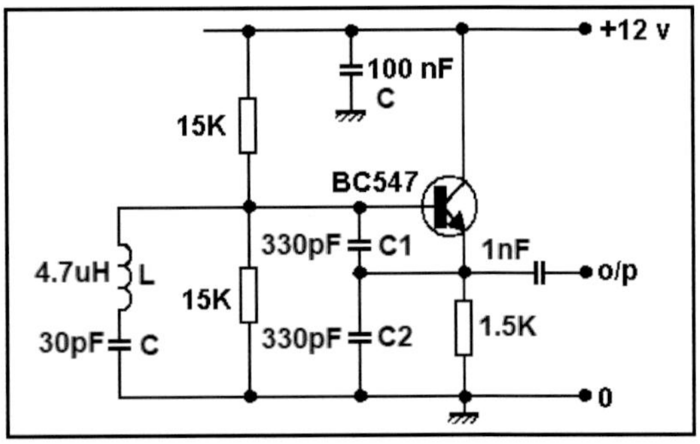

Balanced modulator or mixer using the SA612

The device is a workhorse of QRP designs using double- or single sideband for transmtters and in receivers. Originally it was manufactured for low power mixers in the HF to VHF region. Using a current of 3mA or less, it contains circuitry for multiplying two signals together, one or both of which can be externally supplied, at pin 6 for an oscillator input, and at pins 1 and/ or 2 for generally lower level signals if used as a mixer/modulator. Also, pins 6 and 7 can be used, with external components, to create an internal oscillator, which eliminates the need for an external oscillator signal at pin 6. As detailed data sheets for the device and its connections are readily available on line, a few salient points about practical aspects will be made here.

Use of an external oscillator signal, injected into pin 6, gives more flexibilty over a wide frequency range. Input sinewave levels can be between 200mV peak-to-peak and 750mV peak to peak. Greater than 750mV gives increased distortion when using the device as a modulator, in terms of the modulated output at pins 4 and 5.

The internal oscillator works better for higher frequencies than 10MHz when using crystal or ceramic stabilised oscillator designs, but is good for LC oscillators over the whole specified range of frequencies from zero up to 200 MHz.

It also pays to use as pure a sinewave as possible for the external oscillator input to keep spurious signals as low as possible at the outputs.

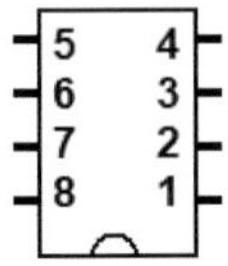

SA612

(top view)

Pin	
1 =	Input 1
2 =	Input 2
3 =	Ground
4 =	Output 1
5 =	Output 2
6 =	Osc 1/buffer
7 =	Osc 2
8 =	V+ (4.5- 8 volts)

The inputs to pins 1 and 2 should be no more than 50mV peak to peak when using the device as a mixer, and no more than 80mV peak to peak when using it as a modulator, to keep output signal distortion low.

Audio amplifier using the LM386.

Again, this is a standard workhorse circuit, although there are many other ICs around which do the same job and are still fairly cheap. Its pin connections are as follows:

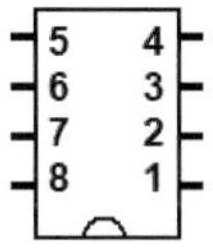

LM386

(top view)

Pin	
1 =	Gain
2 =	- input
3 =	+input
4 =	Ground
5 =	Output
6 =	+Supply
7 =	Bypass
8 =	Gain

It operates from a good range of supply voltages, being +4 to +12 volts up to +5 to +18 volts, depending on the version. The voltage gain is fixed at 20 as a default, but this can be changed using external components. It will drive loads of 4 to 16Ω, with 8Ω being typical. For these loads the total harmonic distortion is below 1% for outputs up to 200 mW, but then it increases rapidly for higher powers or lower load impedances, so it is a low power, personal level audio amplifier, but ideally suitable for QRP work, The quiescent current is typically 4–8mA. A typical arrangement is as below:

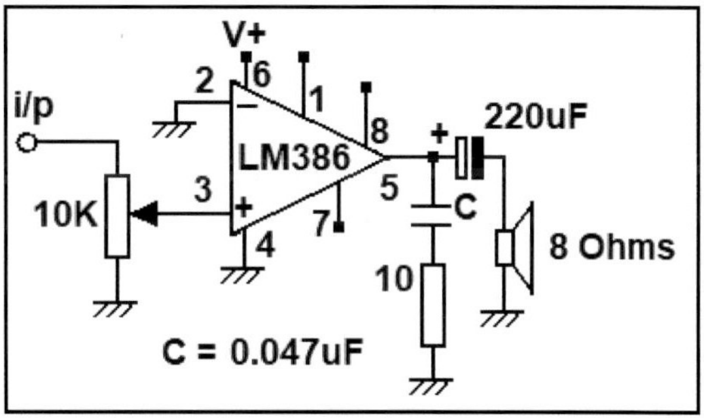

A capacitor connected between pins 1 and 8 of value $10\mu F$ increases the gain from 20 to 200. A resistor in series with that capacitor reduces the gain to a value between 20 and 200, according to its value, as shown in the commercial datasheets for it, readily available online. Note that modules involving the LM386 can be bought online from several sources quite cheaply, which reduces the work involved when making one's own QRP systems, for typically £1-£2.

Chapter 24

Spectrum responsibility and health

In modern times, one of the main concerns of the spectrum regulatory bodies around the world is the use of the RF spectrum by all who are licensed to use it. It is particularly important to keep the spectrum of any transmitted signal tightly within the frequency space allocated for a service. This has consequences in a range of areas of use, and in particular amateur radio, because users have degrees of freedom within the permitted allocations which others, such as in aircraft and marine communications, do not. The problems of spectrum leakage are always more significant when higher powers are being transmitted, as output stages can so easily be overloaded, causing signal distortion and consequently, potentially larger amplitude out-of-band frequency components.

This is one area where QRP is particularly kind to other radio communicators. Even if there are such unwanted frequency components due to RF transmitter output filters not being of the highest quality in terms of out-of-band suppression, then these components are, pro-rata, much lower in power and therefore not contributing RF pollution so widely in the geographic sense. Additionally, even when operating in band, it is clearly bad manners to spread one's spectrum across frequencies used by other in-band users, so one should not overmodulate carriers either.

Another concern of the times is the potential hazards of radio communications in respect of human health. Since the advent of cell phone/mobile phone technologies, there has been much research done on absorption in the human body, indicated as the SAR – the Specific Absorption Rate, which each manufacturer must specify. This is defined as the rate at which energy is absorbed per unit mass by the human body in Watts per kilogram. In the USA this is around 1.6 Watts per kilogram in 1 gram of tissue, whereas in the European Union it's around 2 Watts per kilogram in 10 grams of tissue. The public at large have worried about cell phone powers which can be up to around 600 mW, but are typically 100 mW, and the significance depends on the part of the body exposed, the head being most vulnerable, especially to lower RF frequencies. Consequently, one must take care, when using even 5-Watt handheld transceivers, not to use them too near the head. Loft antennas, or external antennas, for QRP work, are both very good in terms of power density, but those "non-QRP" users who transmit several hundred Watts are even more vulnerable than those who live by Base Transceiver Stations, commonly known as mobile phone masts/cell phone masts. Work done by a former colleague of mine, Professor Peter Excel, in the areas of radio frequency hazards, once told me that, the lower the radio frequency, the greater the potential hazard levels for the human body by radio frequency fields.

One may conclude, therefore, that QRP communications is not only good fun and quite challenging, but also creates much lower radio frequency pollution, and with a considerably-reduced human health risk.

Roger Green M0RJG